EMERGENCY PREPAREDNESS & More
A MANUAL ON FOOD STORAGE AND SURVIVAL

By

Howard Godfrey

Copyright © 2011 Howard Godfrey

All rights reserved. No reproduction of this book, in whole or part in any form whatsoever, is permitted without the expressed, prior, written consent of the author.

ISBN 978-1461196723

Copies of this book may be ordered
From
Howard Godfrey
P.O. Box 3214, Bowman, Ca. 95604 or are
Available from Amazon.com and other retailers

Be sure and visit my blog at
www.Preparednessadvice.com

Dedication

I would like to dedicate this book to my wife Bonnie who has the patience to put up with my hours on the computer. To all the family members and friends who have helped and provided information, Scott Godfrey, Steve Godfrey, FreezeDryGuy, Greg Cline, Ken Sakamoto, Dr Martin Rosengreen, Jim Putnam, Dirk Helder and especially Nancy Godfrey. My many thanks.

TABLE OF CONTENTS

Introduction _____ V

Chapter 1 - How to begin _____ 1

Chapter 2 - Miscellaneous Thoughts _____ 7

Chapter 3 - Water _____ 19

Chapter 4 - Food _____ 42

Chapter 5 - Storage Tips _____ 73

Chapter 6 - Preserving Foods _____ 86

Chapter 7 – Recipes _____ 96

Chapter 8 - Cooking and Heating _____ 119

Chapter 9 – Lights _____ 144

Chapter 10 - Fuels _____ 159

Chapter 11- Medical Supplies _____ 167

Chapter 12 - Sanitation _____ 178

Chapter 13 - 72-hour kits _____ 189

Chapter 14 - Survival _____ 221

Chapter 15 - Improvised Equipment and supplies _____ 252

Chapter 16 - Weapons of mass destruction _____ 262

Chapter 17 – Communication _____ 286

Chapter 18 - Trade and Precious Metals _____ 292

Chapter 19 - Parting Thoughts _____ 304

Reference Section _____ 308

Index _____ 319

Introduction

When I wrote my first book, "Emergency Preparedness the Right Way", I did not intend to write a second. However, since my first book has been on the market, I have received many requests for further information. Because of the success of "Emergency Preparedness the Right Way", I have found myself with access to loads of new information on products and survival tips.

In the past 18 months, I have spent a lot of time lurking on various preparedness blogs as well as researching food storage and other areas of preparedness. The collapse of the Argentina economy, the war in Bosnia and Katrina, have all provided new insights. As a result, this book will go into greater depth than my previous one.

There is no way that any book can cover all the potential problems that are possible in today's world. You only have to read the headlines of any newspaper and you will see articles on terrorism, earthquakes, pandemics or economic collapse.

It is impossible to prepare for every eventuality, but by learning a few skills and storing some basic items, you can improve your chances of surviving a disaster in a reasonably comfortable manner.

This book will not cover every aspect of preparedness. Areas such as firearms, medicine, nuclear, chemical, and biological warfare will only be covered in a limited manner. There are already many good books on these subjects. The reference section in the appendix will refer you to some of the books that I recommend.

I have not included any religious or political opinions in this book nor made any effort to convince you to prepare. If you are reading this, you probably already believe in preparedness and any opinions I express will just waste your time and mine. So let us get on to the subject of preparedness.

Chapter 1 - How to begin

There is an old kid's joke, "How do you eat an elephant?" The answer, "One bite at a time". Emergency preparedness is a lot like this. You do not have to do it all at once. Just make a plan and stick to it.

Planning is one of the most important things you can do in life. Planning will increase your chances for survival during any emergency.

Many of you will be experienced preppers (experienced at preparedness), while some will just be starting. It does not matter where you are at (geographically, age-wise or financially), you should have a plan. I have seen people accumulate large amounts of supplies without a plan. This will often result in wasted time and money. During the last 40 years, I have made many mistakes. You have the chance to learn from them.

As you first read this chapter, you will find many questions you may not be able to answer. For example: How much food or water should you store? By the time you have finished this book, you should know how to find the answers. I recommend that you read the entire book before you start to spend money on food and gear.

I will now take you through a series of steps to assist you in getting started. These steps will help you prepare your plan and start eating that elephant.

The type of supplies you gather and the skills you develop will depend on the type of hazards you may face. The reasons could include economic collapse, earthquakes, tidal waves, nuclear warfare, or terrorism. The following questions will help you decide.

Step 1. Identifying hazards - What are the hazards in your neighborhood or geographic area? Do you live on an earthquake fault? Are you in a flood zone or an area of high fire danger?

Are there major highways or railroad routes (which could cause chemical spills or explosions) near your home?

What about major natural gas lines like the one that blew up in San Bruno California in 2010?

Do you live close to a military base or major government facility? Is there a nuclear power plant near you?

Are your reasons for preparing based on your religious beliefs? Do you fear our government?

Your reasons could include several of the above-mentioned ideas or be as simple as wanting three to five days of supplies for fear of earthquakes or severe weather.

Once you have answered these questions, go to step two.

Step 2. Now that you have a better idea of what you are preparing for, study and determine what supplies and skills will help fulfill your needs.

Are you planning to accumulate a three-day supply or a year's supply of food? I personally believe that you should stock at least one-year's supply of the necessities of life.

What are your most important needs? Normally this will consist of good water, food, heat, or fire for cooking and shelter.

A big part of determining your needs will depend on where you live. Are you in an urban or rural environment? Do you reside in the hot dry desert of the southwest or a cold northern state?

How much do you need to store? In Chapter 4 on food storage, there is information on how to determine the right types and quantities of food for long-term storage.

My personal belief is that during a major disaster; you cannot plan on the government coming to your rescue for at least several days. The local, state, and federal governments as well as the Red Cross all tell us to have at least three days of supplies. However, do not forget the lessons of hurricane Katrina. With a large geographical area of damage, it took the government longer to arrive. In an even larger disaster, it could take weeks or months.

Step 3, Now that you have decided what you are preparing for and for how long, it is time to take that first bite of the elephant. Not very many of us have the financial resources to run out and buy everything we need at once. Therefore, we have to prioritize our purchases. The order in which you prioritize your purchases will depend on many things.

1. Where do you live? What are the weather conditions in your area? Do you have hard winters and lots of rain or snow? Do you live in the desert? Do you need to store large amounts of water or is fuel for heating more important?

2. Are you in an urban or rural environment? How much land do you have?

3. Do you normally plant a vegetable garden? Do you raise crops, or have farm animals? Do you have a well or is surface water available?

4. What are your skills? Can you start a fire without matches? Are you an outdoorsman or a city slicker? Have you had military training? Do you hunt?

5. Are you on prescription medicines? How important are they to you? Think about how you can build up a reserve of your necessary meds. Do you have a good first aid kit?

Now that you have taken your location and skills into consideration, prioritize your needs. Good water is always your first consideration. Second is food or shelter depending on the climatic conditions.

Now take into consideration your skill level and needs. Start to make a list of your basic requirements.
Do not worry about the list being incomplete because it will continue to evolve as your knowledge increases.

STEP 4. Prepare a budget. I strongly recommend that you do not go into debt. Consider eliminating unnecessary expenses. Do you really need cable TV? Plan to buy a few items every month, starting with the highest priority. If you are conscientious about it, you will be surprised by how quickly you will accumulate your supplies.

You can use several tricks to stretch your money. Pay close attention to the sales. Shop online: you can get better prices. Just make sure that you deal with reputable companies. Check the company's reviews online. Buy products that are out of season. For example, purchase cold weather clothing in the middle of summer.

Do not forget about garage sales. I have saved a lot of money buying items like sleeping bags, tents and other camping gear, paying only pennies on the dollar. Remember to haggle so you can get the best prices. You will be surprised at how many discounts you can get just by asking.

Ebay and Craigslist are other good sources. When it comes to Craigslist, you need to be quick and call fast bargains sell quickly. Many of you may be leery about meeting people you do not know. I always like to do trades and purchases in public, well-lit areas like a restaurant or a fast food parking lot.

Chapter 2 - Miscellaneous Thoughts

This chapter consists of a series of miscellaneous thoughts and ideas that you need to take into consideration as you shape your plan.

Firearms - One item that I will not discuss in detail is firearms. There are many excellent books easily available on the subject written by people who know more than I do. I do not intend to try to compete with them. I know that many people put their primary emphasis on arms and ammunition. I do not. My emphasis is on food, water and other necessities. However, it is my personal opinion that everyone who can legally own a gun should.

If you make the decision to obtain firearms, there are a few basic rules you should follow.

- Only own legal firearms.
- Know your state's gun laws.
- Train with your weapons. The more practice the better.
- Practice good gun safety. Keep your firearms in gun safes and out of the reach of children and thieves.
- Do not forget ammunition, extra magazines, cleaning supplies and replacement parts.
- If you are carrying a firearm, do not ever confront anyone unless you are willing to use it. Bluffing or shooting to wound will get you killed.

Martial arts and other forms of self-defense - There are many good forms of martial arts including Karate, Judo, Jujitsu, Boxing, and Wrestling. Whether it is a traditional or a modern school, you can gain skill and self-confidence from attending any of them.

Earthquakes, floods, natural disasters - Make plans for the types of natural disasters most likely to occur near your residence. Check with the local planning agency and learn if you live in a flood zone.

The internet has excellent maps showing the location of earthquake faults. Avoid building in flood plains and near earthquake faults. In some communities, particularly in California, it is impossible to avoid earthquake faults. Consider moving out of the area. Sometimes for family or financial reasons, it is impossible to move.

If you have chosen to remain in an earthquake area there are many things you can do to minimize the potential damage. Make sure that your home meets local building codes. Is your water heater strapped to the wall? Does everyone in your family know how to shut off the water, gas, and electricity?

Keep a pair of shoes near your bed. Cut feet are one of the most common injuries in an earthquake. This is the result of people panicking and running through broken glass and debris to exit buildings.

If you have food storage, have you taken precautions to keep glass containers from breaking? A two to three inch piece of wood running along the front of your shelves will help keep bottles and jars in their place.

Railroads and highways - If possible, live away from main highways and railroad tracks. Trucks and trains transport large amount of hazardous materials. Obtain the latest copy of the Emergency Response Guidebook published by the U.S. Department of Transportation. You can purchase this guidebook over the internet.

The guidebook contains the codes required to read the placards on the sides of the trucks and trains. This permits you to identify what types of hazardous materials are being transported. The guidebook also provides the recommended evacuation distances in case of a spill or accident involving hazardous materials.

Groups - Well-organized groups have many advantages. They offer strength in numbers. They have a greater range of skills and experiences. A larger number of people give you the ability to maintain a watch for security and do all the extra work the situation will require.

Groups can consist of family, friends, churches, neighbors, or just individuals with a common goal. In the process of organizing a group, consider the skills and abilities of the members. Encourage members to develop new skills and learn from each other. Cross training is very important.

In all cases, you face similar problems. Make a Group Emergency Plan. Keep the plan simple and flexible. Allow for changing conditions. Do not become locked into a rigid detailed plan.

Your group may not be together when disaster strikes, so it is important to know how to contact one another and where you will gather. Plan rally points where your group will assemble, both within and outside of your neighborhood.

It may be easier to make a long distance phone call than to call across town, so have an out-of-area contact number. This is because the local system and the long distance systems involve different companies and sometimes one will work when the other won't. If the phones are working, you can leave messages at the out-of-area phone number for other members of the group.

Text messaging is another possibility. In a recent major fire in my area, text messaging worked when the voice systems were overloaded.

You may also want to inquire about emergency plans at daycare centers and schools your children attend. Know what the procedures are for picking up children from school.

Be sure to take into consideration any physical handicaps or other limitations members may have because of age or health.

Knowing whom to trust is an important factor. While I strongly suggest that you do nothing that violates the law, keep your plans to yourselves. The fewer people that know your business, the better off you are.

Retreats - a hideaway in the backcountry. I am not a big believer in retreats unless you can live there on a permanent basis. I know people whose retreats are located several hundred miles from where they normally live and work. It is easy to use all your resources to construct a remote retreat. Then when the time comes to evacuate to your retreat, you may find your route blocked because of martial law, civil disturbance, pandemic, or other causes. Always have a plan and the necessary supplies to shelter in place at your normal residence.

Be sure you have access to sufficient fuel supplies to get you all the way there. If you have to carry fuel inside your vehicles, get good quality approved flammable liquid cans to store it in.

If your plan includes travel to a retreat or distant location, learn several different routes. Get good maps of the areas you will have to cross including topographical maps. Learn how to bypass cities and large communities that are located on your route. The more you can learn about the back roads to your retreat the better. Long after the freeways are blocked, back roads may still be open.

Unless you have a member of your family or a trusted friend living at the retreat full time, it could be looted or occupied by someone else when you arrive.

The average family who lives in the country cannot maintain a 24-hour watch. If you have a prepared retreat, remember that survival is hard work. To maintain a garden, tend livestock; keep a 24-hour lookout, cook, and do all the other chores requires more than a couple of people.

Urban versus rural - If something happens we all want to be living in the country with our family, friends and supplies at a nice, safe, secure place off the grid with water and nice gardens.

Unfortunately, this is a pipe dream for most of us; we will be trapped in the urban environment in which we normally reside.

When the financial collapse occurred in Argentina in the 1990's, city dwellers had to survive in a depressed economy with hyperinflation while living in dangerous and violent areas. Avoiding kidnappings, robberies, muggings, carjacks or being killed became an everyday challenge. There were no heavily armed gangs or mobs of looters trashing homes and businesses in the city. However, there was a major increase in street level crime against property and individuals. Criminals in the city would usually strike fast, grab what they could, and leave before help could arrive.

Those who lived in the city were in a bad situation. However, so were the people who lived in isolated country homes. In some ways, they were in more danger. Organized gangs from the city numbering up to 25 people would drive out to the country. They would find an isolated home that looked like an inviting target.

Even though firearms were available to homeowners, most did not have the manpower to keep a 24 hour watch. The home occupants would then suffer a sudden vicious home invasion. Since the home was isolated with no neighbors the thugs felt safe spending hours or sometimes days stealing everything of value, raping and torturing the victims for hidden valuables.

Living in the country was an advantage when it came to food. Many people had animals, a small orchard, or a garden area. In the cities, food prices went up so high, that people hunted birds and ate street dogs and cats.

The people who avoided the worst of the trouble lived in or near small towns, or close-knit neighborhoods where people knew who belonged and looked out for each other.

We could soon be facing a society that may resemble Brazil, South Africa, Argentina or Russia in the 1990's. These are societies with a corrupt central government, decaying infrastructure, massive economic problems and extremely high levels of violent crime.

General guidelines - for urban survival: Keep your preparations secret. Do not brag to neighbors, friends, etc. Know your neighborhood and neighbors; make friends. Know the location of your local water sources; for example, fountains, ponds, streams and reservoirs. In apartment buildings, the water systems including piping, water heaters and toilet tanks can hold a substantial amount of water. In multi-story buildings, there are large volumes of water in the fire sprinkler systems. In cold climates, some fire sprinkler systems contain anti-freeze.

Do not run generators or show lots of light at night. If you have a generator and lights, put up blackout curtains. You can improvise them from blankets or other heavy materials. Avoid attracting attention to yourself. Watch for like-minded people with whom you can ally.

There is a saying in the military, "If I can find your MRE (Meals Ready to Eat) trash, I can find your base". This also applies if your neighbor sees your trash. He knows what you are eating. What are you doing with your trash? Can your neighbors see where you are throwing your food wrappers? Burn your paper waste and bury what is noncombustible. Can your neighbor smell food cooking?

Rural area - One of the biggest problems for people who live in rural areas will be refugees from the cities. They may be hungry and desperate.

Hundreds, possibly thousands of people, may flee to the countryside seeking refuge. A good portion of these people will have little if any outdoor camping experience. Depending on the time of the year and the ground cover, their attempts at camping will result in wildfires. This group will include a small percentage of anarchists and pyromaniacs who may deliberately set fires. When you make your plans, consider the strong possibility of major unchecked wildfires that involve hundreds of thousands of acres.

In discussing preparedness with various groups, I always find someone who thinks they can go off into the woods and survive off the country. Invariably, this person has little hunting or survival experience and their opinion is based on survival books they have read. Most people who try this will lack the experience and equipment needed to succeed and just become another refugee.

Grocery stores in the U.S. are mostly on a "Just in Time" delivery schedule. This means that with computers they maintain an accurate inventory and as an item is sold, an order is made for its replacement. Deliveries are made daily and very little reserve stock is kept in the back room. What you see on the shelves is almost the entire stock. Forget any ideas of stocking up at the last minute.

Vehicles - Keep your truck or car in good shape. This means your maintenance should be current and keep your gas tank at least ½ full at all times. Your tires should be serviceable, and properly inflated. Always have at least one good spare tire in the vehicle. Make sure you have a jack and lug wrench with you. A tool kit and a few spare parts are good ideas. Include duct tape, electrical tape, plastic cable ties, and wire. It is amazing what you can fix with these few items.

I recommend that you keep a "Get Home Bag" in your vehicle. This is a bag with the necessary supplies to get you home if your vehicle is out of commission. In Chapter 13, I have listed the items contained in mine. Yours will vary depending on your travel distances and weather conditions.

Your body- A lot of us maintain our car, stock the house with food, and forget about our bodies. In an emergency, the most important tool you have is your body. Are your inoculations current? Go to the dentist and get your teeth cleaned and checked. A bad tooth may cost you lots of pain in a long-term emergency and can affect your overall health.

How about your medications? You need to stock up on any prescription drugs you are taking. Talk to your doctor and see if he will help you. Consider purchasing your meds through the internet. Know your local drug laws to avoid committing a felony.

Are you in good enough shape to carry your 72-hour (bug out bag) pack for a few miles? Exercise regularly. You may never be a great athlete, but lose these extra pounds and get in the best shape you can.

Psychology of Survival or Will You Eat a Rat?

"Will you eat a rat?" is a legitimate question. If your answer is no, you are not mentally prepared to survive. Rats are a delicacy in parts of the world. One of the first ideas the military teaches about surviving in a prisoner of war camp is no matter how bad the food, never miss a meal. In a real survival situation when you are short of food, you have to eat anything and everything. A meal missed is calories you may never get back.

During World War II when rationing was tight and food was short, many Europeans ate roof hares (a euphemism for cats) and horses. The trick is to make up your mind ahead of time that you are a survivor.

Early in the Second World War, the Merchant Marine noticed that the average age of survivors from sinking ships were men in their forties and fifties. Research showed that the younger, stronger men were giving up faster. This is why the Military pushes you to exceed your limits. Most of us have never really found out what we are capable of doing.

Once you experience the lack of things you take for granted, like food, medicine, and clean water, your priorities change. My father was raised during the Great Depression. He sometimes went to bed hungry. I can still see how this affects him. He always wants to have extra food around; it is a priority for him.

Regardless of what equipment you have hidden away, it all comes down to having the emotional stability, the determination and the knowledge to use it. You will die or become a burden on your friends or family if you become an emotional basket case.

I have a strong belief in God and feel that this gives me the spiritual strength to face adversity. Whatever system of beliefs or principles you follow be sure that they are strong enough to sustain you. Remember, you are mentally and physically stronger than you think you are.

Chapter 3 - Water

There is an old saying, "three minutes without air, three days without water, and three weeks without food. This statement is not quite accurate. In some areas, you may not last three days without water. Good clean uncontaminated water is always a priority. In hot desert climates like the southwestern United States, you may require two gallons or more a day. I live in a more moderate area, but I always plan on at least one gallon a day per person. The amount and method of water storage you require depends on the following considerations:

Water sources – If you are lucky, you will have a good underground source of water. I know people who have wells or natural springs on their property that test pure. If not, there are many other options, which we will discuss in this chapter.

Wells - In some rural areas there are still old-fashioned hand dug wells. If you have one on your property, they require maintenance. Hand dug wells are usually quite shallow, typically less than 25 feet deep. Today it is very difficult to find sanitary drinking water in a shallow well because surface runoff and shallow subsurface water enter the water supply. A solid stone or concrete wall around the top of the well will serve two purposes. It will divert surface water and keep children from falling in the well. Some of the older, hand dug wells are lined with stone or concrete. If these liners are kept in good, watertight condition, they will keep shallow subsurface water from entering the well.

If you have a full cover over the hand-dug well, a pump will help reduce chances of contamination significantly. This can be hand operated or an electric pump. In addition, a full cover will keep wildlife out of the well. In rural areas where pump maintenance and repair can be a real problem, large diameter wells are often the best solution to water supply problems. Pumps can be installed while leaving an access way through which water can be drawn by rope and bucket if the pump should break down.

If you intend to hand dig your own well, it is best to do it during the part of the year that is the driest.

Cisterns - Most cisterns are designed to catch and store rainwater. The difference between a well and a cistern is that the cistern has a waterproof lining and is a storage area. To keep a clean water supply, cisterns must be kept clean. You should inspect them regularly. Make sure that the liners are watertight. You can empty a cistern and clean them with an appropriate dilution of chlorine. Be sure to rinse them well.

Cisterns can be very effective. When I was in Bermuda, I saw them in use. In Bermuda, because of their water shortage, the law requires that every house collect 80 percent of the water that falls on its roof. They use white tile roofs and rain gutters to collect the water and divert it into their cisterns.

An improvised cistern can be made with ¾ inch plywood, plastic tarps and 2x4's. A simple box measuring 4x4x8 feet and lined with plastic will hold approximately 955 gallons of water. One inch of rain from a 2000 sq ft roof will equal about 1200 gallons. Be sure and use the 2x4's to reinforce the boxes.

Deep wells - If you are on a deep well, you need alternate sources of power to operate the well pump. If you don't have an alternate source, a simple method for getting water out of a deep well without electricity is shown in the following diagram.

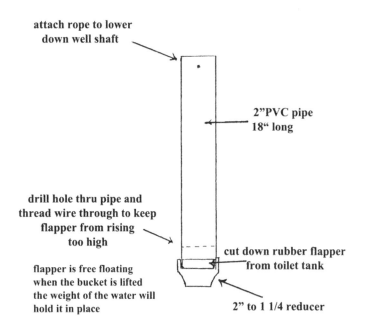

Attach a rope to the PVC pipe and lower the bucket down the 4 to 6 inch well shaft, and let it sink into the water. The rubber flapper will act like a foot valve and rise up against the wires when it hits the water. This will allow the water to enter the pipe. When you start to pull it up, the weight of the water will push the rubber flapper down against the reducer and seal the bottom of the bucket.

Streams and surface water are not a safe source of drinking water. I live in a semi-rural area in which a lot of gold mining occurred. The miners used mercury in large quantities and water from the mines may be dangerous to your health. Be familiar with the history of what has occurred in your neighborhood.

Do not trust water from springs or mountain streams. The statement has been made that running water is safe to drink. Do not believe it. There are many old wives' tales about beds of sand and gravel purifying water. The space between grains of sand is more than enough to let microscopic giardia lamblia parasites pass between them. Trust me, you do not want giardia. A young healthy, friend of mine had it and lost 40 pounds in about three weeks!

Giardia and Cryptosporidium are two waterborne diseases often found in swimming pools and hot tubs as well as in rural settings. They are spread by human and animal feces and are hard to destroy.

During the Civil War more men died from diseases caused by unsanitary conditions than from combat.

Read the section on water purification later in this chapter.

Urban environment - If you reside in a city or urban area, you are probably dependent on a municipal water system. In times of turmoil, they will not be reliable. Be aware of other sources of water in your neighborhood. For instance, rivers, streams, swimming pools, or fountains can be useful sources of water.

As many old neighborhoods change, wells are capped and abandoned. Someone who has lived there for years may remember the location of one.

The plumbing system in high rises, office buildings, apartments and other large structures contain large amounts of water. Become familiar with the building you have access to and learn how to isolate the water system from the municipal supply. Once the system has been isolated from the municipal water system, you can drain water as needed. It is best if the building system is isolated from the municipal system before the municipal system becomes contaminated. If you encounter contaminated water, use the purification methods described in the remainder of this chapter.

Contamination - Are there natural or manmade contaminates in the water? One should ask this question before drinking. Contaminates could include animal or human waste as well as chemicals. I have a stream across the street from my house. In good times, it is contaminated with animal waste. In an emergency, a large percentage of the population would not know how to deal with waste products and would contribute to the contamination. Washing clothes, dishes and bathing in available surface water are some examples. Assume that any water other than from a deep well or a cistern you maintain is contaminated.

Storage - Ok, so how much water should you store? Figure a minimum of one gallon a day per person. This only includes drinking, cooking, and very basic hygiene, such as washing dishes and brushing teeth.

Therefore, if you have a family of five and you decide you needed water for two weeks, 5 times 14 equals 70 gallons. In my opinion, this is the absolute minimum. Storing 70 gallons of water is a lot easier than it sounds. You probably have 40 gallons in your water heater and several gallons in the toilet tanks. If you plan to use the water from the toilet tanks, do not add any cleaning materials or disinfectants to the tank water. Do not use the water from the toilet bowl.

At the first sign of an emergency, shut off any connections to a municipal water system. The water in your pipes and water heater should be safe to drink. If you do not isolate your plumbing from the municipal water system, the water in your system may become contaminated. Be sure to turn off the heat before draining a water heater.

How do you store water without spending lots of money? You can go to the local sporting goods store, and buy several 5-gallon containers. Of course, these are going to run you from $10 to $20 dollars each. If you have the money this is all right, but there are cheaper ways.

A friend of mine saves his two-liter soda bottles, washes them out, and fills them with fresh water. He then throws them in the crawl space under his home. Two-liter soda bottles make excellent storage containers. These are strong, light and designed to hold liquids. Every day you throw away high quality storage containers, such as juice and water bottles.

Beware of plastic milk containers. They have a tendency to break down and the lids do not seal well. In the past, many sources have recommended bleach bottles. However, the current manufacturers of bleach advise against the use of bleach bottles for water storage.

Food grade plastic containers that are marked with the recycling number 1 or PETE or PET are safe for use. Do not use containers that have been used to store nonfood items. Plastic bottles are permeable and should not be stored near flammable liquids, pesticides, or other chemicals. They will pick up tastes and odors of chemicals and flammable liquids stored in close proximity.

Water has been stored in everything from waterbeds to canning jars. Beware of waterbeds, because they are not made of food grade plastic. They may contain dangerous chemicals. There are pros and cons to all methods. Fifty-five gallon barrels or other large containers are too heavy to carry and often require you to siphon or pump the water out. Inexpensive hand pumps are available. You can improvise a siphon hose.

Fifteen and thirty gallon containers are lighter and easier to move. They are often used for the delivery of syrups to various food manufacturers and restaurants. I have found excellent containers of various sizes in surplus yards. If you know a restaurant or bakery owner, check with them to see how they dispose of their containers.

Five-gallon containers are easier to carry. They weigh about 50 lbs each. If you have to walk to a water source, you can run a pole through the handle of one or more five-gallon containers and two people can share the weight. Water weighs 8.35 lbs. per gallon.

Glass jars are an effective way to store water. They are non-permeable and will not pickup bad tastes or odors. Remember to protect glass; it is easily broken in earthquakes or other disasters. A great place to store water bottles is in the available free space in your freezer. This makes your freezer more efficient and the ice helps to keep the food cold in a power failure. If you store water in the freezer, remember to leave room for the ice to expand.

Storing the water in various types of containers is easy. Be sure the containers are clean by thoroughly washing all the containers prior to filling. A sanitizing solution can be prepared by mixing 1 teaspoon of liquid chlorine bleach (5 to 6% sodium hypochlorite) to

1 quart of water. Check, the strength of your bleach as some companies now are only using 3.75% sodium hypochlorite. This requires you to use a stronger solution. Use only household bleach without thickeners, scents, or additives.

If your water is from a chlorinated municipal water system, do not add Chlorox or any other chemicals to the water. Store your containers out of the sun light. Rotate the water once a year. When you go to use the water, it should be fine. However, I intend to treat mine prior to use just to be safe.

Now that you have information on how much water you require a day and ideas on how to store it, determine how much water your family needs. Personally, I store a minimum of a month supply for my family. When that runs out, I plan to use the water from the stream across the street. Remember, as I mentioned earlier, this is a contaminated stream. This water requires treatment before using it for drinking, cooking, dishwashing, or brushing teeth.

The following are some of the more common and readily available methods for treating water. First, if the water is muddy or turbid, let it set until the particles settle to the bottom, then drain the clear liquid off the top. You can also use a fine cloth or chamois to filter the water prior to treating it. A quick funnel can be manufactured by cutting the bottom out of a two-liter soda bottle. Turn it upside down and fill the neck with a fine weave clean cloth or chamois. Do not use the synthetic chamois that are now on the market. Fill the bottle with water through the bottom and let it filter through the cloth and out the neck of the bottle. This will not purify the water, but will take out the large particles and extend the life of your water filters.

Chlorine dioxide tablets manufactured by Aquamira, Micropur, and Portable Aqua, kill bacteria, viruses, and cysts, including Giardia and Cryptosporidium. Chlorine dioxide is a well-established disinfectant. Chlorine dioxide is iodine and chlorine free. Check the expiration date on the package. Dioxide tablets have a shelf life of four to five years. Aquamira and Micropur meet the EPA guidelines for Microbiological Water Purifiers. The U.S. Military is currently switching from Iodine tablets to Chlorine Dioxide tablets. The downside to chlorine dioxide tablets is that they take several hours to work, so you have to plan ahead.

Iodine tablets - One tablet added to a quart releases eight ppm (parts per million) of iodine. Two tablets are used for turbid water. Wait 15 minutes after adding tablets, 30 minutes if the water is cold. The shelf life of the tablets is 3-5 years. Deterioration is evidenced by a change in the tablet color; metallic gray is acceptable, light yellow double the dose, and reddish brown discard. The formation of a precipitate is acceptable. If there is starch in the water from potatoes, corn, or rice the water will turn a blue color. The blue-colored water is harmless and is acceptable to drink. Iodine tablets are not completely effective against Giardia and Cryptosporidium. Iodine can have adverse health effects on some people.

Warning - Pregnant or nursing women or persons with thyroid problems should not drink water disinfected with Iodine.

Chlorox or chlorine bleach - Common household Chlorox or chlorine bleach with 5 to 6% sodium hypochlorite may be used to disinfect water in the following amounts: Four drops per quart gives 10 ppm (parts per million) in clear water. This amount should be increased to eight drops in turbid water. Sixteen drops will provide 10 ppm per gallon of clear water. You should be able to get a slight odor of Chlorox after the waters sits for 15 minutes. If not, add more Chlorox. Chlorox or chlorinated bleach loses it strength with time. After one year on the shelf, Chlorox will have lost 50% of its strength, so double the dose. Remember to buy plain, unscented bleach with no thickeners or additives.

Warning - Chlorine or iodine will not reliably kill Giardia and Cryptosporidium. At colder temperatures, doubling or tripling the wait time will improve your chances. SODIS (Solar water disinfection), pasteurization, boiling, chlorine dioxide tablets, and good water filters are more reliable.

Solar water disinfection (SODIS) - Pour clean water into clear food-grade PET (Polyethylene Terephtalate) bottles and expose to sunlight for a minimum of six hours if the sky is bright or less than 50% cloudy. If the sky is 50 to 100 % overcast, the container needs to be exposed to the sun for two consecutive days. This method uses solar radiation and heat to destroy pathogenic microorganisms, which cause waterborne diseases.

SODIS requires a full bottle of water that is clear enough to read through. Plastic bottles made from PET (PolyEthylene Terephtalate) or clear glass bottles are preferred. Avoid the use of bottles made of PVC (PollyVinylchloride). PVC bottles contain UV stabilizers, which blocks the sun's radiation. PVC bottles often have a slight bluish color. When burned, PVC plastic gives off a pungent smelling smoke. The smell of burning PET is sweet. Heavily scratched or old bottles should be replaced due to a reduction of UV transmittance, which will reduce the efficiency of SODIS. The following recycling table shows the symbol and unicode. One of these symbols is on the bottom of most bottles.

Plastic Identification Code Type of plastic polymer Properties Common Packaging Applications:

Polyethylene terephthalate (PET, PETE) Clarity, strength, toughness, a barrier to gas and moisture. Soft drink, water, and salad dressing bottles; peanut butter and jam jars

PE-HD

High-density polyethylene (HDPE) Stiffness, strength, toughness, resistance to moisture, permeability to gas. Water pipes, Hula-Hoop (children's game) rings, Milk, juice and water bottles; the occasional shampoo /toiletry bottle.

PVC

Polyvinyl chloride (PVC) Versatility, clarity, ease of blending, strength, toughness. Juice bottles; cling films; PVC piping.

PE-LD

Low-density polyethylene (LDPE) Ease of processing, strength, toughness, flexibility, ease of sealing, a barrier to moisture. Frozen food bags; squeezable bottles, e.g. honey, mustard; cling films; flexible container lids.

PP

Polypropylene (PP) Strength, toughness, resistance to heat, chemicals, grease, and oil, versatile, barrier to moisture. Reusable, microwaveable ware; kitchenware; yogurt containers; margarine tubs; microwaveable disposable take-away containers; and disposable cups and plates.

PS

Polystyrene (PS) Versatility, clarity, easily formed Egg cartons; packing peanuts; disposable cups, plates, trays and cutlery; disposable take-away containers.

O

Other (often **polycarbonate** or **ABS**). Dependent on polymers or combination of polymers Beverage bottles; baby milk bottles; electronic casing.

Laying the bottles of water on sheets of corrugated sheet metal or on a roof is the preferred method. This helps to increase the temperature. Attempts have been made to make large trays using window glass to purify water. Beware of this method; much of today's window glass has UV inhibitors or tints that block the UV rays.

The SODIS method was developed by the Swiss Federal Institute for Environmental Science and Technology and is recognized by the United States Center for Disease Control.

SODIS does not work well during periods of rainfall. Harvesting rainfall is suggested during rainy spells.

Water Pasteurization Indicator (WAPI) is a simple thermometer that indicates when water has reached pasteurization temperature and is safe to consume. Pasteurization destroys all microorganisms that cause diseases from drinking contaminated water. The WAPI consists of a small polycarbonate tube that contains a soy wax that melts when water is heated enough to be pasteurized (65°C/149°F). Heating water to 149°F kills E. coli, rotaviruses, giardia, and the hepatitis A virus.

This saves fuel by eliminating the need to boil water to ensure that the pasteurization temperature has been reached.

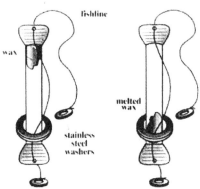

Instruction on how to use a WAPI- The WAPI can be used with most fuels, including wood, gas, kerosene, charcoal, but it was originally designed to work with solar ovens.

Fill your container (do not use plastic) with water, place over heat or in solar oven. Place WAPI in water with the wax filled end up. To do this you have to slide the WAPI to the end of the string so that the wax end is furthest from the washer.

This puts the wax end nearest the surface of the water. When the water reaches the correct temperature, the wax melts and falls to the bottom of the WAPI. This shows that the water has been pasteurized and is ready to drink.

Remember pasteurization does not remove dangerous chemicals. Pasteurization is not the same as sterilization and the WAPI should not be used for medical or food canning. See the reference section for suppliers that carry the WAPI.

Backpack filters - There are many poor quality filters currently on the market as well as many good ones. Make sure the filter you purchase will filter protozoa, bacteria, and viruses down to 0.5 microns. Make sure your filter is certified to EPA Guide Standards for microbiological purifiers against bacteria, cysts, and viruses. Three reputable brands are First Need, Katadyn, and Aquamira. The Frontier Pro by Aquamira is currently my first choice for a small filter to use in a 72-hour kit or Get Home Bag. The filter is small, light, inexpensive, and will purify about 50 gallons of water. Most backpack filters are limited in the quantity of water they will treat, often as little as 20 to 40 gallons. New models that have recently come on the market treat much larger amounts. Check the specifications and buy the best you can afford. This is not the place to economize. Learn to use your filters. They take a little practice. Do not forget to store extra replacement filters.

Gravity flow filters - These are simple, easy to use and can treat thousands of gallons of clean pre-filtered water. Merely pour the pre-filtered water in the top container and wait for it to flow though the filters into the bottom container. It requires no electricity or special knowledge. The most recommended brands are the plastic American Berkefeld, the stainless steel British Berkefeld and the stainless steel Aqua Rain. Some of these filters can treat up to a gallon an hour. These units have been laboratory tested and will remove dangerous organisms such as protozoan cysts (Cryptosporidium, Giardia lamblia) and microscopic bacteria (E. coli, Salmonella typhimurium, etc).

For a permanent residence, a good gravity flow filter is superior and much more convenient. Gravity flow filters do not require chemicals. Since they operate by gravity, no pumping is required. They purify your water while you are free to do other things and they have a one to two gallon storage reservoir. A tap on the bottom reservoir is very convenient.

Aqua Rain gravity flow filter

American Berkefeld gravity flow filter

Rainwater from your roof can be collected in a 55-gallon drum. Rig a downspout from your rain gutters to a clean barrel. Place a screen over the top of the barrel to prevent mosquitoes from breeding. Do not assume that the rainwater will be fit to drink without treatment. However when I was a child in South Africa, we drank untreated rainwater directly from the roof for a couple of years without ill effects. The water was collected in a large galvanized tank with the faucet about 18 inches from the bottom. This allowed sediment to collect at the bottom of the tank. Remember, you have birds and insects living on the roof and their feces end up in the tank.

SteriPEN is a unique product that utilizes ultraviolet (UV) light technology to purify water, destroying more than 99.9 percent of bacteria, viruses, and protozoan cysts such as giardia and cryptosporidium.

The method has been in use for over one hundred years, and is used to purify drinking water by some of the largest cities in the world,

SteriPEN's are available in several models including a new version called the Sidewinder, which requires no batteries. It will purify 8,000 liters of water. You have to hand crank 90 seconds for each liter of water you purify.

The remaining models of the SteriPEN require four AA lithium or metal hydride batteries. Some models are available with a solar charging case that includes a double-cell rechargeable CR123 battery and provides secure storage for the purifier.

Sidewinder by SteriPEN

Sweetwater Microfilter Purifier System, I recently had the chance to use one of these filters. It was simple and quick to assemble or disassemble.

The unit has a lever handle that gives you a four to one advantage when pumping. My wife has arthritis in her hands, but she was able to pump water easily. The rating which says the filter will pump one liter a minute is very conservative.

The specifications shown below are for the Sweetwater Microfilter Purifier system.
- Effective against protozoa - Yes
- Effective against bacteria - Yes
- Effective against viruses - Yes
- Effective against particulate -Yes
- Effective against chemicals/toxins - Yes
- Weight 11 oz/320 g
- Width 2 in/5 cm
- Length 7.5 in/19 cm
- Country of Origin: Made in Seattle, USA

The only disappointing thing was that the filter is only rated for 750 liters. While you do have the ability to clean the filter with the brush that is provided, I have not yet determined how much this will extend the filter life. Always try to use the cleanest possible water. If the water is turbid, fill a container and let the heavy particles settle to the bottom. This will help extend the life of the filter.

This is a filter that I would have no problem putting in my 72-hour kit or taking backpacking.

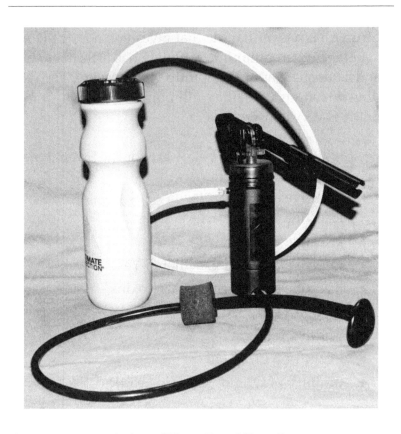

Sweetwater Microfilter Purifier System

Platypus Gravityworks Pump Free Water System - This is a gravity flow system that will filter 4 liters of water in two and a half minutes. It is simple to use and I found it very user friendly.

In the above picture, you can see both the clean and dirty water bags and the filter. The filter is good for 1500 liters. The filter is effective against both giardia and cryptosporidium.

Chapter 4 - Food

The type of food that you store is a very personal matter. It depends on a combination of things: your normal diet, your budget, and your storage space.

First, store food you and your family will consume. People always say that if they are hungry enough they will eat anything. Unfortunately, experience has shown this to be a fallacy. Monotonous diets cause appetite fatigue. This is not too much of a problem for healthy adults, but the elderly and children may refuse to eat enough to stay healthy.

How much food should you store? This depends on your goals and beliefs. Why are you storing food and for how long do you think you will need it? My suggested plan consists of storing a three months supply of food that will be rotated through your normal diet. This will be backed up by enough long-term food storage for at least a year.

Food is like the cash in your wallet; it disappears quickly. In an emergency you want your family to stay healthy and in good spirits. Do not underestimate the amount of food this will take. Being hungry can lead to malnutrition, sickness, and poor morale. Many of the prepackaged programs are based on 1500 - 2000 calories a day. I feel this is inadequate. For example, a man doing hard physical labor can require 4000 calories or more per day.

How do you accumulate this quantity of food without blowing your budget? First rule: do not go in debt. It is like the saying, How do you walk a hundred miles, you do it one-step at a time. Purchase something extra every month. Work at it steadily whenever you get a little extra cash.

Three Months Supply

The three months supply consists of canned food (wet pack), packaged foods, boxed cereals, and frozen food. In other words, the three months supply will contain the foods we consume in our normal diet, with a few exceptions. Fresh produce and meats may be unavailable. Items that require refrigeration can be included, if you have a backup plan in case of power failure. This could be an alternate power source to run your refrigerators/freezers or preserving your food by canning, salting or drying. If you live in a cold climate, you may be able to store food outdoors. A friend in Wyoming often stores meat in a small lean-to behind his house during the winter. Just remember to protect your cache from both two and four legged animals.

To help you determine the quantities of food you need to store, I have included a table showing the food rationing in England during World War II.

The average standard rations during the World War II are as follows. Quantities are per week unless otherwise stated.

Food Rations

Item	Maximum level	Minimum level	Rations (April 1945)
Bacon and Ham	8 oz (227 g)	4 oz (113 g)	4 oz (113 g)
Sugar	16 oz (454 g)	8 oz (227 g)	8 oz (227 g)
Loose Tea	4 oz (113 g)	2 oz (57 g)	2 oz (57 g)
Meat	1lb3oz	1lb3oz	1lb3oz
Cheese	8 oz (227 g)	1 oz (28 g)	2 oz (57 g)
			Vegetarians were allowed an extra 3 oz (85 g) cheese
Preserves	1 lb (0.45 kg) per month 2 lb (0.91 kg) marmalade	8 oz (227 g) per month	2 lb (0.91 kg) marmalade *or* 1 lb (0.45 kg) preserve *or* 1 lb (0.45 kg) sugar
Butter	8 oz (227 g)	2 oz (57 g)	2 oz (57 g)
Margarine	12 oz (340 g)	4 oz (113 g)	4 oz (113 g)
Lard	3 oz (85 g)	2 oz (57 g)	2 oz (57 g)
Candy	16 oz (454 g) per month	8 oz (227 g) per month	12 oz (340 g) per month

Notes:

1 egg per week or 1 packet (makes 12 "eggs") of egg powder per month (vegetarians were allowed two *eggs*) Arrangements were made for vegetarians so that their rations of meat were substituted by other goods.

Milk was supplied at three imperial pints (a little larger than a US pint) each week for expectant mothers and children under five. Each consumer got one tin of milk powder (equal to 8 imperial pints every 8 weeks.

There were no ration restrictions on fruit, potatoes, fish, or vegetables. Most types of fruit and vegetables were hard to find, especially onions. Bread was not rationed, but no white flour was available. People were encouraged to dig up their lawns and flower beds to create "victory gardens", and grow their own vegetables.

The good effect of rationing was that by the end of the war the poor people of Britain had never been so healthy!

People ate larger amounts of vegetables and grains and less protein. A typical World War 2 recipe for "Woolton pie" is included in Chapter 7.

Now that you have some idea of how little you can get by on, I want you to make a list of the foods you actually use for one week. Use this list to extrapolate how much food you will need for three months.

Canned food (also known as wet packed). A frequently asked question is: how long is the shelf life of canned foods? Shelf life varies depending on the product. Foods with a high acid content have the shortest shelf life. Tomato products are a good example.

A majority of cans have at least one or two dates on them: A best if used-by date and a manufactured date that is often in code. The codes showing manufactured dates vary by manufacturer and usually include the time and place of canning. The best if used by date is not an expiration date. Expiration dates can be determined by calling the toll free number that most manufacturers provide.

If correctly stored in moderate temperatures (below 70 degrees F), the majority of canned foods will last at least two years past the best if used by date. Most canned products retain their safety and nutritional values long beyond these dates.

I have personally eaten canned goods that were seven to eight years old past their best if used by date. Prior to eating any out-of-date canned goods, check the cans carefully for bulges or leaks. If you open a can and it spits at you or shows a release of gas pressure, do not eat the foods. Try not to even touch the contents; there is a good chance that the can contains botulism. In theory, as long as the cans are sound and not bulging the food should be safe to eat.

Canned foods 100 years old have been tested and found to be safe to eat, but I would not recommend it. Time always causes a loss of nutrition and deterioration in taste. In my own storage, I have had to throw out cans that were bulged or leaking.

Rotate your short-term foods on a regular schedule. When you purchase canned products always check the best if used by date to determine which products have the longest remaining shelf life. In most grocery stores, they place the products with the shortest shelf life at the front. Always check the products in back. Do not buy dented or damaged cans. Beware of clearance houses and dollar stores as they often have products that are reaching the end of their shelf life.

Items that you may want to include:

Canned Products
- Tuna
- Any canned fish
- Meats
- Soups
- Vegetables
- Fruit
- Condensed Milk
- Chili
- Fruit or Vegetable Juice
- Spam
- Corn Beef
- Canned Stew

Package Items
- Packaged cereal
- Spaghetti
- Macaroni
- Legumes (split peas, beans)
- Instant Potatoes

- Rice
- Soup Mixes
- Cake Mixes
- Pancake Mix
- Flour
- Sugar
- Honey
- Crackers

Spices and Condiments
(There are some general rules for storing herbs and spices. When possible, purchase whole spices. Keep them cool and in dark places).
- Salt
- Pepper
- Cayenne Pepper
- Bouillon cubes (beef and chicken)
- Misc. Spices
- Ketchup
- Mustard
- Syrup
- Jellies and Jams

Oils and Fats
- Olive
- Canola Oil
- Corn Oil
- Canned Shortening

The above list is just a short list of suggested grocery items. Your list should be considerably longer. Remember to rotate your food.

Oil has a shelf life of one to two years before turning rancid. Oxygen is eight times more soluble in fats than in water. This causes oxidation and is the primary reason oil turns rancid. The more polyunsaturated oil is, the faster it will go rancid. Vegetable oils (such as safflower, corn, sunflower, flax, soy and cottonseed oils), nuts and seeds contain the highest amount of polyunsaturated oils. Rancidity may not be readily apparent to you because vegetable oils have to become several times more rancid than animal fats before our noses can detect it.

Exposure to oxygen, light and heat all contribute to rancidity. If possible, refrigerate your stored oil, particularly after it has been opened. Try to buy your oils in opaque, airtight containers. If you purchase it in plastic, at your first opportunity, transfer it to a gas impermeable glass or metal container that can be sealed airtight. If you have a means of vacuum sealing the storage container, use it. This removes most of the air remaining inside, taking much of the oxygen with it.

Transparent glass and plastic containers should be stored in the dark, such as in a cupboard or box. Regardless of the type of storage container, you should store it at as cool a temperature as possible and rotate it as fast as is practical. Oils and fats with preservatives added by the manufacturer will have a greater shelf life than those without them, provided they are fresh when purchased.

Crisco shortening or hydrogenated shortening in metal or metal lined cans have the longest shelf life of any fats (except for freeze-dried or canned butter). The shelf life of canned shortening that have preservatives added is estimated to be from 8 to 10 years. Some authorities claim that if kept cool, the shelf life of freeze-dried butter is almost indefinite.

I have seen reports that coconut oil if properly stored will last up to 5 years. These reports are from producer's web sites. I have not been able to find any definitive studies on this issue.

With liquid oils, be sure to watch the expiration dates and rotate them as needed. Extra virgin olive oil is reputed to have a slightly longer shelf life, because it is the first squeezing and therefore purer.

Rancid fats have been suspected of causing increases in arteriosclerosis, heart disease, and cancer. Whenever possible, keep oil away from light and oxygen and store in a refrigerator or cool place.

For long-term storage of fats, shortening powder (dehydrated shortening) is available through many of the dehydrated food companies. It is claimed to have a twenty-year shelf life.

Yeast - Active dry yeast should have a shelf life of approximately 1 year at 70 degrees. Keeping it refrigerated should extend the storage life to around 3 to 5 years. Freezing it will extend the storage life, but you should proof a sample every year to be sure it is still active. You proof yeast by mixing a small quantity of yeast with an equal amount of sugar. Add the mixture to warm water (105-115 degrees). Active yeast will begin to expand and become bubbly within five to ten minutes. If the reaction takes longer, you can still use the yeast, but you will need to use more. Yeast that shows no reaction should be discarded.

Vinegar is used for preserving food, as a condiment, salad dressing, medicinally, as a disinfectant and as a cleaner. There are numerous types of vinegars on the market, but for the purposes of this book, we will only consider white distilled vinegar and apple cider vinegar. White distilled is not true vinegar but is actually diluted distilled acetic acid. It will store almost indefinitely if tightly sealed in a glass or plastic bottle with a plastic lid. The acid will destroy enamel-coated metal caps over time. It works well for pickling and most other uses.

Apple cider vinegar is sold in two types: one is a cider flavored distilled acetic acid, the other is a true cider vinegar fermented from hard cider. Fermented apple cider vinegar will occasionally form a cloudy substance. This is not harmful and can be filtered out prior to use or consumed. The cloudy substance is called Mother of Vinegar and can be used to make more vinegar. If the vinegar starts to smell bad, throw it away.

Long-Term Storage

Long-term storage foods are defined as foods that are considered safe and nutritious for extended periods, up to 30 years.

An example of long-term food items are freeze- dried, dehydrated, and staple food items such as grains and beans with a low moisture content (10% or less) that can be stored for 20 to 30 years.

Oscar A. Pike, the lead researcher at BYU Long-term Food Storage Research team and Chair of the Department of Nutrition, Dietetics, and Food Science, answered the following questions about food storage:

Question: "How long will stored foods stay good?"

Answer: "There is a wide range in the shelf life of dried foods, depending on the specific commodity and its original quality, storage temperature, and so on. Some commodities should be used within a couple of years, like salad oil and dried eggs. However, many dried foods — packaged to remove oxygen and kept at room temperature or below — will store well for 20–30 years or more. In our studies, taste testers evaluated aroma, flavor, texture, and overall acceptability of dried foods. Wheat and rice were very acceptable after 30-plus years of storage; beans, dried apples, macaroni, potato flakes, and oats up to 30 years; nonfat dry milk up to about 20 years."

Question: "Do foods that old retain their nutritional value?"

Answer: "There is a loss of nutrients over time, but there is sufficient nutritional value to justify storing dry foods long-term. In a survival situation, you need calories to stay alive, and stored foods provide calories. Vitamin C is another important nutrient and fortunately, vitamin C tablets retain a high percentage of their potency for more than 20 years."

How much should you store? Here are some guidelines on the amount of food you need for one adult per year.

For an adult you should store a minimum of the following.

A combined 400 lbs of wheat, rice, other grains, beans and other legumes

60 lbs of sugar or honey

75 lbs of low fat powdered milk

5 lbs salt

If you intend to survive only on the above, it would provide enough calories, but barely meet nutritional standards. After a few days, the diet would become rapidly monotonous and the roughage would be hard on your system.
.

If you live in the country, supplementing your diet with fresh vegetables, eggs, milk, and some fresh meat would improve your nutrition. However, considering that the majority of us live in an urban environment, this is not practical. You need to store a mixture of dehydrated and freeze-dried foods to supplement your nutrition and provide variety. They should include vegetables and meats.

Planning what you store may be one of the most important decisions that you are ever required to make. Too many people buy the food package provided by food storage dealers. They are often a one-size fits all package. Others work off a list provided by friends or obtained from books. The goal of this book is to teach you to think for yourself. You need a food package that fits your family, taking into consideration health problems and ages.

Let us discuss some common foods.

Wheat should be part of the 400 lbs of mixed grains and legumes that you store per person. Remember that about 40% of the population has a wheat allergy. Most allergies are mild and not of any consequence in our normal diet. If wheat were suddenly to become a large part of our diet, many individuals would have severe symptoms.

Personally, I store wheat because it is nutritious and has good protein value. The flour is versatile, and it is easy to crack for cereal. Sprouted wheat provides fresh greens high in vitamin C.

There are several diverse types of wheat that have different uses:

Duram wheat is the hardest of all U.S. wheat. It is planted in the spring. Duram contains a high percentage of protein and is used for making macaroni, spaghetti, and other pasta products.

Hard red spring wheat contains the highest protein content of all U.S. grown wheat. It has high gluten content and makes excellent bread.

Hard red winter wheat is commonly used for bread and all-purpose flour. It has a good percentage of protein and gluten. It is probably the most common wheat for home storage because it comes closest to meeting all the requirements.

Hard white wheat is fairly new to the United States. It has a milder sweeter taste than red wheat. It is generally used for yeast breads, bulgur, and hard rolls

Soft red winter wheat has a low to medium percentage of protein and gluten. It is used to make pastries, cakes and crackers.

Soft white wheat has a low to medium percentage of protein and gluten. It is used for pastries and cakes.

Kept in # 10 cans with oxygen absorbers, wheat should last for 30 years (see Storage Tips Chapter 5). If your wheat is over 30 years old do not be in a rush to get rid of it. Check it and if it appears to be good try to sprout some. If it sprouts, it is as good as new. If it fails to sprout, it will have lost many of its nutritional values but it is still good for calories.

White and whole flour have approximately 1/3 the storage life of whole wheat.

Spelt is similar to wheat in appearance. However, spelt has a tougher shell than wheat, which helps protect the nutrients in it. Spelt has a somewhat nuttier and slightly sweeter flavor than whole-wheat flour. Spelt contains more protein than wheat (17%), and the protein in spelt is easier to digest. This means that some people who are allergic to wheat may be able to tolerate spelt. Spelt has gluten, just like wheat, so spelt is not suitable for a gluten-free diet. Spelt flour can replace whole-wheat flour or whole grain flour in recipes for breads and pasta.

Since spelt is a hard grain and is related to wheat, it should have similar storage characteristics. No tests have been done to confirm this that I am aware of. However, I know people who have successfully stored it for many years.

Hulled Barley is commonly used to add thickness to soups and stews. It has low gluten content so it does not make good leaven bread. It makes good flat bread and porridge. Barley contains eight **essential amino acids**. Hermetically sealed with oxygen absorbers under good conditions hulled barley will store for 25 years.

Quinoa can be used like rice. It should be washed prior to use or it will have a bitter taste. Hermetically sealed with oxygen absorbers in good conditions quinoa will store for 25 years.

Buckwheat, Flax, Kamut, and Millet are all hard grains. When stored in a cool dry place, hermetically sealed with oxygen absorbers they will store for 30 years.

Corn - When corn was first introduced into non- Native American farming, it was generally welcomed with enthusiasm for its productivity. However, a widespread problem of malnutrition soon arose wherever corn was introduced as a staple. This was a mystery since these types of malnutrition were not normally seen among the Native Americans to whom corn was the principal staple food.

It was eventually discovered that the Native Americans learned long ago to add alkali — in the form of wood ashes among North Americans and lime (calcium carbonate) among South and Central Americans — to corn meal. This liberates the B-vitamin niacin, the lack of which was the underlying cause of the condition known as pellagra.

Besides the lack of niacin, pellagra was also characterized by protein deficiency, the result of a lack of two key amino acids in corn, lysine, and tryptophan. The Native Americans had learned to balance their consumption of corn with beans and other protein sources, such as meat and fish, in order to acquire the complete range of amino acids for normal protein synthesis.

Corn, despite its limitations, is still an excellent storage food. The best variety to store is yellow flint or dent corn. They are low moisture if properly dried. They make good polenta meal and flour. Popcorn should not be ground in most mills due to its extreme hardness. Several mills such as The Family Grain Mill and the Back T Basic Mill recommend that their mills not be used for popcorn.

Oats - Oats are mainly thought of as a bland breakfast food in the United States, but in Scotland and Ireland, they were considered a staple. They are an excellent source of iron, dietary fiber, and thiamin. One of the benefits of consuming oats is antioxidants, which are believed to protect the circulatory system from diseases such as arteriosclerosis, which affects the arterial blood vessels.

Uses of oats include cereals, a thickener for soups and stews, filler in meat loaf and casseroles, pancakes and baking. Regular and quick rolled oats are the most commonly stored and can last 30 years if properly packaged.

Studies by Brigham Young University stated the following about the storage of oats. "Flavor and texture was significantly affected by oxygen level.

Although loss of quality occurred during storage, the level of acceptability indicates that including rolled oats in long-term storage is a viable option".

Rice - It is probably the most consumed food in the world. White rice is the only form that is viable for long-term storage. If stored properly white rice will store almost indefinitely. Tests by Brigham Young University shows it to be edible and nutritious after 30 years of storage.

Rice is an excellent source of complex carbohydrates. All eight of the essential amino acids are contained in white rice. During the milling process, white rice looses approximately 10% of its protein, 70% of its minerals, and 85% of its fat. In addition, thiamin, niacin, and iron are lost during this process. Any rice sold as enriched has had thiamin, niacin, and iron added after milling.

Brown rice is more nutritious, but due to its fat content should not be stored for any longer than 6 months, since it will turn rancid.

Legumes - Various cultures worldwide have used a combination of grains and legumes for centuries as the staple of their diet. This has proved itself as the basis for a healthy diet.

This variety of food is one of highest in protein for non-animal foods containing between 20%-35%. It consists primarily of beans, peas, and lentils. Legumes by themselves are not a complete protein, but when combined with other grains (corn, rice) become a complete protein. Beans contain all essential amino acids, except methionine. Methionine is found in corn, rice, or meat. Beans are an excellent source of fiber, starch, minerals and some vitamins.

Legumes include Beans of all types, a partial list includes
- Adzuki
- Black eye peas
- Black turtle beans
- Chickpeas or garbanzo beans
- Great northern
- Kidney beans
- Lentils
- Lima beans
- Pinto beans
- Soybeans
- White beans

Long-term storage information on beans has not been readily available in the past. However, recent tests by Brigham Young University have established that black, white and pinto beans will store for over 30 years and still be acceptable for use.

All beans, even when they are correctly stored (see Chapter 5 Storage Tips), become harder over time and take more time to cook.

All dried beans, except lentils and split peas, require soaking in water for rehydration.

A method to help soften beans and speed up the cooking is as follows: First, sort and rinse the beans. Bring three cups of water to boil for each cup of beans. Add the beans to the boiling water and bring to a rolling boil for two minutes. Take the beans off the stove. Next, add 3/8 teaspoon of baking soda (sodium bicarbonate) for each cup of beans cover and soak for 1 hour or more. Extra baking soda may be required for older beans. Drain and rinse the beans thoroughly. Cover the beans with water and bring to a boil, then reduce the heat and simmer 1-2 hours or until tender. Do not add salt or other ingredients until the beans have softened adequately.

Bean flour - If beans reach a point that they will no longer rehydrate, grind them into bean flour. Bean flour provides protein, some carbs and a lot of fiber, as well as additional nutrients. Bean flour can be substituted in many applications that use wheat and other traditional flours. An added benefit is bean flours can be eaten by people with celiac disease, whereas wheat products (as well as rye, barley, and to a lesser extent, oats) cause trouble.

Split peas and lentils - During the 18th and 19th centuries, split peas were used by both the British and American Navies as a shipboard food with great success. Storage was in wooden casks, by today's standards, a very inadequate storage method.

Peas and lentils are packed with so much fiber, protein, and other nutrients that the USDA recommends that legumes be consumed as both a meat and vegetable selection.

Jordan S. Chapman, Brigham Young University, Provo, UT in a presentation made in July 30, 2007, stated the following: "Ten samples of split peas representing 5 retail brands packaged in size No.10 cans and stored at room temperature were obtained from donors. Two fresh samples of split peas were purchased as controls. Samples ranged in age from one to 34 years. Can headspace, oxygen, can seam integrity, and split pea water activity and color were evaluated. A 52-member consumer panel evaluated the samples, prepared as split pea soup, for appearance, aroma, texture, flavor, and overall acceptability using a 9-point hedonic scale."

"Acceptance for use in everyday and emergency situations was also determined. Can headspace oxygen ranged from 0.19 to 20.1%. All can seams were determined to be satisfactory. Hedonic scores for texture declined over time, corresponding with increasing hardness of the peas. All samples had an acceptance in an emergency situation of over 75%. Results indicate split pea quality declines over time, but the product maintains sufficient sensory acceptance to be considered for use in applications requiring long-term storage."

When you purchase legumes or grains, the best choices are pre-cleaned products. Most obtained from food dealers will be pre-cleaned. If you are buying bulk from a producer or distributor, you may be buying field-run. This has not been cleaned and may be quite dirty. Know where your food comes from. Avoid animal feed products that may be subjected to fumigants that are forbidden for human consumption.

Pasta stores longer than flour. Correctly packaged with oxygen absorbers and stored in a cool dry place it should last 20 years

Nonfat dry milk - A strong source of calcium, protein, and vitamin A. Regular nonfat dried milk is my personal choice for long-term storage. Its lack of fat keeps it from turning rancid. Tests at BYU indicate that properly stored in #10 cans with an oxygen absorber, it should last 20 years. The reason for recommending nonfat regular over instant milk is that it is more compressed, needs less storage space and is usually cheaper. The disadvantage is that it is a little harder to prepare.

If you purchase your milk in bulk, repack it into smaller containers. If left in the original bags its storage life will be greatly shortened and it will attract insects and rodents.

Tests conducted by the Department of Nutrition, Dietetics and Food Science, Brigham Young University, revealed the following information:

"Twenty samples of regular and instant nonfat dry milk (representing 9 brands) stored up to 29 years at ambient conditions were obtained from 14 sources. Samples were evaluated for headspace oxygen, can seam quality, water activity, solubility index, sensory quality (50-member consumer panel using a 9-point hedonic score), and nutritional value."

"Headspace oxygen ranged from 0.05%-20.9 percentage, which was related to the efficacy of the oxygen removal treatment (nitrogen, carbon dioxide or oxygen absorbers). Only six samples had less than 2% headspace oxygen. However, a 23-year old sample with low oxygen was not significantly different from fresh samples."

"Though there is some decline in quality over time, it appears possible to retain palatability and nutritional value in nonfat dry milk during long-term storage by using adequate packaging and storage conditions."

Morning Moo, Swiss Maid, and several other milk products are whey-based. This means that most of the milk protein has been removed. Whey based milks are not a good choice if you are counting on the powdered milk for protein, especially for growing children. Most recipes for powder milk call for instant powdered milk. If you have non-instant use the chart below to determine quantities.

Powdered Milk Reconstitution Chart

For:	Instant Dry Milk	Non-Instant Dry Milk	Water (Approx.)
1 gallon milk	5 1/3 cups	3 cups	15 cups
1 qt. milk	1 1/3 cups	3/4 cup	3 3/4 cups
1 pint milk	2/3 cup	3/8 cup	1 3/4 cups + 2 Tbsp.
1 cup milk	1/3 cup	3 Tbsp	Scant cup
1/2 cup milk	3 Tbsp.	1 1/2 Tbsp	1/2 cup
1/3 cup milk	2 Tbsp + 1 tsp	1 Tbsp	1/3 cup
1/4 cup milk	1 1/2 Tbsp	2 tsp	1/4 cup

Baking powder Tests conducted by the Department of Nutrition, Dietetics and Food Science, Brigham Young University, revealed the following information: "Baking powder is widely used to leaven baked products. The industry standard for baking powder shelf life is eighteen to twenty-four months, but little information has been available on baking powder functionality when stored beyond this time."

"The objective of this research was to determine the effect of long-term storage on baking powder functionality. Six samples of double-acting baking powder in original commercial packaging were obtained from donors and two fresh samples were purchased. Samples ranged in age from 0.25-29 years and were stored in cool (15-25 °C) and dry conditions. Biscuits were made following standardized procedures and measured for height, diameter, and surface crumb color."

"Under optimal storage conditions, it appears that baking powder retains its functionality as a leavening agent for many years and can be included in applications requiring long-term food storage."

Baking soda stored in an airtight container and kept dry will store almost indefinitely. In its original non-airtight cardboard box it will keep for about 18 months and absorb any odor it is exposed too. Baking soda will leaven bread. It can be used to make hominy.

Salt - I personally store what many people would consider an excessive amount of salt. It is cheap, stores indefinitely if protected from moisture and can be retained in its original packaging. Salt was used as a preservative prior to refrigeration (salt fish, salt pork, salt beef, etc.) Because of its many uses as a food preservative and its low cost, I store at least 100 lbs of salt.

Canning salt should be used as a preservative. This contains no additives such as iodide or anti-clumping agents. Canning salt may form clumps when exposed to moisture, but it does not hurt the salt. Just break up the clumps.

Sugar - Granular white and brown sugars have an indefinite storage life if stored in insect and moisture-proof containers.

Honey - Sweeter than sugar, it stores well. I prefer honey to sugar for personal use, although I store both. Honey has a tendency to crystallize with age. This is not a problem. In England when I was a child, it came this way from the store. I still like crystallized honey spread on bread. If you do not like it crystallized merely heat the container up in a large pan of hot water and it turns liquid.

Do not store honey in unlined metal containers. With age, it takes on a metallic taste and turns black. You will not eat it.

Warning - Do not give honey to infants under the age of one as it can cause them to suffer from botulism.

Textured vegetable protein, or TVP, is a soy product, low in fat and high in fiber and protein. Vegetarians and vegans use TVP to increase their protein intake and to mimic the texture of meat in a variety of dishes. Some emergency preparedness organizations recommend that people keep TVP, to have a readily available source of protein in a disaster.

TVP is made with defatted soy flour, which is a by-product of the manufacturing process used to make soybean oil. The soy flour is mixed with water, cooked, extruded, and then dried. As it dries, the textured vegetable protein loses the bulk of its weight and turns into small flakes, which resemble breakfast cereal or perhaps dried vegetables.

TVP is a controversial subject. Many experts question the use of soybeans in our diet. There are strong indicators that they inhibit the body's digestion of some vitamins and proteins. Many flavored TVP products are reported to be high in sodium and MSG and some contain partially hydrogenated oils. I do not store it, preferring freeze-dried meats.

Additional low-moisture foods that store well over the long-term include macaroni, onion flakes, potato flakes (not pearls) and spaghetti.

Dehydrated and Freeze-Dried Foods.

One of the problems with long-term low-moisture food is that many of them become deficient in vitamins A, C, B12, and calcium over time. The addition of dehydrated fruits and vegetables and freeze-dried meats can remedy this.

Vitamins A and C can be found in canned or bottled fruits and vegetables as well as in some fruit drink mixes. During dehydration of fruits and vegetables, most vitamin C is destroyed, but some vitamin A remains. Good sources of vitamin A include canned pumpkin and dehydrated carrots.

Vitamin B12 comes from animal products. It is found in canned meats, freeze-dried meats, and jerky.

Calcium comes mainly from dairy products such as powdered milk, hot cocoa mix, and pudding mix (containing dried milk).

Vitamin E is derived from fats and oils. Nuts such as sunflower seeds and almonds are a good source of vitamin E.

Dehydrated Food

Dehydrated foods consist primarily of fruits and vegetables that have had 98% of their moisture removed by drying. This process reduces the moisture in them to levels that inhibit the microbial growth that causes them to rot. They can be stored for extended periods if properly packed (Chapter 5 Storage Tips). Tests indicate that they retain their nutritional value and taste for extended periods of time. Dehydration also reduces weight, which makes them a great backpacking and survival food. I have listed some examples below.

Dehydrated carrots - An excellent source of vitamin A. Tests presented by Stephanie R. Bartholomew, Brigham Young University showed the following: carrots packed in a #10 can with an oxygen absorber would be found acceptable after being stored for 25 years.

I have mentioned carrots because it is a product that has undergone extensive testing. Peas, corn, onions, apples, etc., can be stored long-term if properly packed.

Understand that dehydration is merely removing the moisture, or old-fashioned drying. You can do it at home with a food dehydrator or even in your backyard using the sun.

Onions – A food I would strongly encourage you to add to your storage. Other than salt, they do more to improve the taste of your food than anything else I know. They provide nutrition and fiber to your diet.

Mixed vegetables - I have a friend who buys frozen mixed vegetables, green beans, peas or corn when on sales. He dries them in an electric food dehydrator making sure that they are completely dried. Without doing anything to them other than placing them in a zip lock plastic bag, he has successfully kept them for over five years and they were still edible. The nutritional contents have not been tested. I would recommend canning them with an oxygen absorber if you intend to place them in long-term storage. This will help preserve nutrition.

Commercial dehydrated products are available through numerous suppliers. Most of them are good reputable products. However, read the labels carefully. Some companies are now selling dehydrated food from China. Personally, I prefer to stay with products made in America. Many of these products will store for twenty to thirty years. Read Chapter 5 Storage Tips for more information.

Freeze-Drying

Freeze-drying is the process of drying foods by placing frozen foods in a vacuum at absolute pressures that permit ice to change directly to vapor. In other words, this can be compared to freezer burn, but is so fast that the food shows no effect.

The type of products typically freeze-dried includes meats, fishes, shrimp, instant coffee, vegetables, and fruits.

Advantages
- Little thermal damage
- Excellent flavors
- Good vitamin retention
- Rehydrates rapidly
- Little shrinkage
- Long product storage life, 25 years—if suitably packed

Disadvantages
- High production costs
- Not good for everything
- Rapid deterioration unless products are packaged and stored properly
- Cannot be done at home

Warning - If you intend to store meat, cheeses, or similar products for long-term storage, it is strongly recommend that you choose freeze-dried. When you purchase freeze-dried or dehydrated products, ask about the residual oxygen. This is the amount of oxygen left in the can when it is ready for sale. There are numerous inferior products on the market. Anything you buy should be in good quality metal cans and have no more than 5% residual oxygen and preferable no more than 2%. If your supplier cannot provide you with this information and back it up with laboratory tests, don't buy them.

Chinese freeze dried products are currently being imported in the United States. I recommend you avoid them.

Chapter 5 - Storage Tips

The first rule of food storage can be summed up in the acronym HALT. It stands for the four enemies of good food storage. These are Humidity, Air (oxygen), Light, and Temperature. This is the basis of all food storage.

The following are some facts you need to understand:

Oxygen absorbers - Placed in a non-permeable container, they absorb the available oxygen. Oxygen is one of the main causes of food spoilage. It allows pests and molds to grow in your food. Eliminating oxygen in your long-term food storage containers ensures the longest possible shelf life.

Oxygen absorbers work with a simple chemical process. They contain iron powder and salt which reacts with the oxygen in the air causing the iron powder to rust. When all the iron powder has rusted, the oxygen absorbers are finished and the absorbing action stops.

The reason that the container needs to be non-permeable is that when the oxygen is absorbed, the container maintains a partial vacuum. The number of oxygen absorbers required depends on the size of the container. A 500 CC oxygen absorber is more than adequate for a # 10 can. For a five or six gallon container, you should use three 500 CC oxygen absorbers. If you are using plastic buckets and oxygen absorbers, the buckets need to be lined with a non-permeable barrier such as a Mylar bag.

There are two common types of oxygen absorbers on the market. One is beige and turns a blue-green when expended. The second is bright pink and turns a dark blue-green when expended. When you use oxygen absorbers, remember to always read the instructions that come with them. Do not eat the contents of the oxygen absorber packets.

Glass bottles - For many years glass bottles have been used to store dried foods successfully. Oxygen absorbers work well in glass bottles because glass is an excellent vapor barrier. Make sure that glass bottles are protected from breakage in case of an earthquake or other disaster.

Plastic food grade containers - This includes all sizes of plastic buckets. Containers of food grade quality are manufactured from polycarbonate, polyester, or polyethylene. All sizes vary in characteristics in terms of density, permeability, and strength. Only buckets manufactured from food grade plastics that have a gasket in the lid seal should be used to store food.

There seems to be a great deal of confusion on what constitutes a food grade bucket. Many people are under the mistaken impression that if the recycling number is a 2 that the bucket is automatically food grade. This is an error; it merely means that the bucket is made from a type of plastic that can be food grade.

The final determination is based on the chemicals used to release the molds and provide color.

If you buy a bucket in a store, make sure it is food grade. Some people say it does not matter if they are food grade if you use a Mylar bag liner. I have not found any definitive information on this. However, I feel it is smart to get food grade buckets. You never know what you may have to use them for in the future.

An important fact to understand is that most plastic is permeable; in other words, it breathes. Unless lined with Mylar bags you cannot count on plastic buckets to protect food from oxygen. Do not store food in plastic bags or buckets near gasoline, kerosene or other chemicals - they may pick up the taste and odor. This includes food grade plastics.

Typically, wheat and beans have been stored in plastic buckets without Mylar bags. Based on the observation of others and myself, I feel this method works. I have used wheat that has been stored this way for 20 years. However based, on recent tests, wheat and beans stored only in plastic buckets, while edible, are not as nutritious as those stored in Mylar bags or metal cans and protected from oxygen. All my new grains and legumes are now stored in #10 cans with oxygen absorbers or plastic buckets lined with Mylar bags and oxygen absorbers.

Food grade bucket with Gamma lid

For everyday use, I recommend you keep several food grade buckets with gamma lids. The gamma lids consist of a ring that snaps onto the bucket and a lid that screws into the ring. This makes for a bug tight seal that is easy to open. You will find it very convenient for the food you use every day.

PETE plastic containers are a type of clear plastic bottles containing food products that are sold in the grocery stores. The marking PETE or PET will be on the bottom next to the recycling emblem. The recycling number will be a one. They are a good oxygen barrier and can be used with oxygen absorbers to store bulk dry foods.

Tests conducted by Brigham Young University on used PETE soda bottles showed they would keep the oxygen percentage low enough to disinfest the grain stored in them. Experiments were carried out to determine how long oxygen absorber packets could keep the oxygen level below 1%, the level required to disinfest grain. It was determined that the low moisture grains could be stored at least a year without any appreciable increase in oxygen. The bottles need to have screw on lids with plastic, not paper lid seals. Rodents can chew through PETE bottles.

The following types of containers do not work well for long-term food storage:
- Translucent plastic containers such as milk bottles.
- Containers with snap on lids.
- Any container that has contained a non-food product (should not be used period).

Killing insect infestations in buckets - Oxygen deprivation has also been shown to be an effective method of killing insect infestations when the oxygen content is held below 1% for at least 12 days.

Five-gallon food grade buckets are often used for dry foods like wheat, rice, and beans. Brigham Young University conducted experiments to determine how long the oxygen content of 5 gal buckets filled with wheat could be held below 1% by the use of oxygen absorber packets. It was possible to use oxygen absorbers to reduce the oxygen level below 1% for 12 days. However because plastic is permeable this treatment was shown to be an unreliable method, because the oxygen levels in the samples exceed 1% too frequently. Therefore, it is not recommended that consumers use this method to eliminate insect infestations in grains. If you are going to use a plastic bucket without Mylar bags, dry ice is recommended for killing insect infestations.

Dry ice treatment

This method has been used successfully for storing grains and beans in plastic buckets for many years. It will prevent insect infestation. Treatment methods that depend on the absence of oxygen to kill insects, such as oxygen absorbers, are not effective in plastic buckets. Plastic buckets breathe and without a Mylar bag liner, oxygen will slowly re-enter to the bucket. Dry ice kills the insects but does not provide the contents with any protection from oxygen.

Dry ice treatment instructions.

1: Use approximately one ounce of dry ice per gallon capacity. A five-gallon capacity would require five ounces of dry ice. Do not use metal containers because of the possibility of an explosion caused by excessive buildup of pressure.

2: Always wear gloves when working with dry ice.

3: Using a clean towel, wipe the frost crystals off the dry ice.

4: Place the dry ice in the center of the bottom of the empty container.

5: Fill the bucket with the grain or beans to within an inch of the top.

6: Place the lid on the bucket; snap it partly shut. Leave space for the carbon dioxide gas to escape. This gas is formed as the dry ice changes from a solid to a gas.

7: Before sealing the bucket, allow the dry ice to completely change to a gas. Feel the bottom of the bucket. If it is very cold, dry ice is still present.

8: After the lid is sealed, monitor the bucket for a few minutes. If you notice any bulges, loosen the lid to relieve the pressure.

Warning: Do not stack plastic buckets over three high. With age, many buckets become brittle and collapse if stacked too high. If the buckets are stacked, periodically check the buckets to ensure that the lids have not broken from the weight.

How much food will a five-gallon bucket hold?

- Dried beans approximately 35-40 lbs
- Lentils or split peas approximately 40 lbs
- White sugar or salt approximately 35 lbs
- Wheat approximately 35 lbs
- Rice approximately 33 lbs

Food grade Mylar bags - Mylar bags create an oxygen barrier to protect food during extended long-term storage. Mylar bags can be clear or metalized. Mylar bags are used by some of the better food companies to line their plastic buckets to create an oxygen barrier.

Metalized Mylar bags can be used independently for food storage. They can be sealed with either a standard bag sealer or a common household iron. Just be sure to tug on the seal to make sure it is tight. Mylar bags come in varying sizes and thicknesses. The most common is 4 millimeters.

Aluminum coated plastic pouches are frequently mistaken for Mylar bags. They possess similar characteristics and can be used interchangeably with Mylar bags. Seal them only with an approved pouch sealer. Irons will not provide an adequate seal, especially for powdered products such as flour or dry milk.

Advantages of Mylar bags and aluminum coated plastic pouches.

- Can hold a vacuum
- Inexpensive
- Lightweight Requires little special equipment o use Can be reused
- You can store many empty bags in a small pace

Disadvantages
- They are not rodent proof. If rodents are a problem, store, the pouches in a rodent proof container.
- Odd shaped, awkward to store
- They are subject to being easily torn

Metal cans - Cans are available in several sizes, the most common being Numbers 10, 2-1/2 and 303. These numbers indicate the standard sizes for each can.

- A #10 can generally holds 13 cups.
- A #303 can holds 2 cups.
- A #2-1/2 container holds 3-1/2 cups – just 1/2 cup shy of a quart.

The #10 size cans are often mistakenly called "gallon cans" but they actually hold 5/6 of a gallon. If you have access to a can sealer and oxygen absorbers, cans are the best method of storing dry products for long-term storage in my opinion.

Advantages
- Rodent proof
- Waterproof
- Can hold a vacuum
- Easy to move and store

Disadvantages

- Cannot be resealed
- More expensive than Mylar bags Subject to rust
- If you are purchasing your food products already packed, look for the following:

Any grains or legumes sold in plastic buckets should have Mylar liners.

Freeze-dried products should have the oxygen removed by one of the following methods:

- Vacuum packed
- Nitrogen filled (preferred method)
- Oxygen absorber

The residual oxygen should be no more than 5%, preferably less than 2%. If your supplier cannot provide you with this information and back it up with lab tests, run and find another dealer.

A reputable supplier will not have any problem providing you with information about the residual oxygen. One supplier that I have dealt with for many years describes his packaging in the following manner: "The cans are nitrogen packed, replacing air with nitrogen. Each can is coated with protective enamel, including the lid. This enamel helps protect the can from the deteriorating elements of oxygen and moisture. The contents of the can are protected for many years. Our foods will have the longest shelf life available...in excess of 30 years! As long as the can is not opened or punctured." He has lab reports documenting that his products contain less than 2% oxygen. See <u>Freeze Dry Guy</u> in the list of references at end of this book.

Many people have access to can sealers and Mylar bag sealers. There is nothing wrong with packaging your own products. Be careful of the following:
Since you are probably using oxygen absorbers, you have to realize that they start to work as soon as they are exposed to air. Open and shut the bags as rapidly as possible. When you purchase your oxygen absorbers, ask about the bag clips that are available to help keep the oxygen absorbers fresh.

Check the seals on both the cans and the Mylar bags. The can seals should be flat and tight. Mylar bag seals should be continuous and not pull apart when subject to light stretching.

Remember to date your cans and mark what is in them. I know people who have a pile of unmarked mystery cans in their storage. When you open your cans, use a resealable plastic lid to protect the contents from moisture. The products are best if used within a month or two after opening.

At the beginning of the chapter the acronym HALT was mentioned. If you follow the preceding packaging methods, you have taken care of the air, light, an oxygen portions.

The ideal storage area is a dry, spacious basement that maintains a temperature of 70 degrees or less all year round. If you are lucky enough to have these conditions, all you have to do is arrange your shelves so that the products have air space between the floor and exterior walls.

As for most of us, we have to cram food into all the nooks and crannies of our homes. Normally garages are too hot in the summer to make good storage areas. If possible, you want a storage area where the temperatures do not exceed 70 degrees.

I cannot give you a simple answer on how to store your food. A wise man once said that if you have to use your storage, you would wish you had kept the furniture in the garage and the food in the house!

You have to use your imagination. I have seen bookcases made with five-gallon buckets and box springs set on number 10 cans. Many newer homes have pantries. If your home has a wood floor with a crawl space underneath, you can use an old trick to help cool your pantry. Install a vent in the floor and one in your attic to create an upward airflow from the cool crawl space. Under floor crawl spaces can make excellent storage areas if you do not have a moisture problem.

Managing your Food Supply.

Keep track of the dates of your perishable foods such as wet pack, and rotate them often to insure having the freshest product when you need it.

Date your cans or cases. Use a marking pen and write on the cans.

Watch for rodent or insect infestation in bulk packed items.

Check for rancidity. When in doubt, open and inspect. I would rather open a few items unnecessarily than find that they had gone bad when I needed them most.

If you don't want the hassle of managing your food storage (most people, including me, are too lazy), then buy the best quality freeze-dried and dehydrated foods and grains in #10 or #2-1/2 size cans from someone you trust.

Maintain an inventory of your long-term food storage. As part of the inventory, list the location of the stored item. As your food storage grows, you will find that you store things throughout your house, and if you don't list the location, it is easy to lose track of things.

Chapter 6 - Preserving Foods

"Almost half of the food produced in the world goes to waste due to lack of a means to preserve it." - Dr. Dieter Seifert

This is the most important paragraph in this chapter. There are two books I strongly suggest that you purchase: The Complete Guide to Home Canning by the United States Department of Agriculture and So Easy to Preserve by the Cooperative Extension, The University of Georgia. These books are inexpensive and cover canning, drying, pickling jellying, and freezing in great detail. If you follow their methods, you will not go wrong.

These books contain information on how to preserve meat, vegetables, and fruits. They are by far the best I have seen.

Old Fashion Storage Methods

Canning- not really old fashioned, but not as common as it once was. I am not going to tell you how to can. Instead, I will refer you to the books mentioned in the previous section. After having completed quite a bit of research on canning methods, I have found that a lot of the information on the internet is extremely unreliable.

Open-kettle canning and the processing of freshly filled jars in conventional ovens, microwave ovens, and dishwashers are not recommended. Open-kettle canning is the method of heating the product in a separate pan and then pouring it into jars. A hot lid and ring were then placed on the jar and tightened. This completed the processing. This is not a safe method of canning.

Pressure canning is the only safe way to can vegetables, poultry, meat, and seafood. These are low acid foods and require heating to 240 degrees Fahrenheit.

Just remember when you store glass jars to protect them from breakage.

Solar canning -"If I could use my solar oven for only one thing, it would be for canning," says Eleanor Shimeall author of **Eleanor's Solar Cookbook**. "Processing food in a solar cooker is much simpler than conventional methods. It takes about 10 minutes of preparation time starting from fresh fruit to placing the jar in the cooker. An added bonus is that the heat of the sun does not destroy the color of the fruit."

You can use the standard canning jars. The "terminal sterilization" method works well for solar. You start with clean glass jars and lids; add fruit, sugar, or salt, and water to fill up to the neck of the jar. Space must be left in the jar for expansion during cooking. Screw lids onto the jars as tightly as you can and place the jars into the solar cooker. Let the contents of the jars come to a boil. Once some of the contents of the jar boil out from under the lid, the jar and contents are sterilized.

Lids with a rubber ring are designed to release steam during canning. After the jars have boiled and some of the contents have run out, remove the jars from the oven. Do not remove the jars until they have run over. Wipe clean and allow to cool down slowly. As the jar cools, the lid is pulled downward to form a vacuum seal. Check each cool jar by gently pressing down on the lid making sure it does not move up and down. If a jar does not seal, you may reprocess it.

Solar canning is safe for acidic foods only. Acidic foods are fruits, or their juice, jams, or jellies. DO NOT CAN MEATS OR VEGETABLES IN YOUR SOLAR COOKER! Do not even add a sprig of parsley as botulism can grow on any non-acidic food!

If you decide to attempt this method of canning, be very careful to be sure you are reaching the required temperatures.

How to keep food cool - without ice or refrigeration. An iceless refrigerator that uses water for cooling is still in use in many third world countries. It is simple and cheap to build. The instructions for constructing one are in Chapter 15.

If you have clean running water or a spring, or cistern, you can place food in the water to cool. Protect the food from being contaminated by polluted water.

Drying

The purpose of drying is to preserve food by lowering the amount of water or moisture in the food to a point where microbial growth and chemical reactions cannot destroy the food during storage.

Because drying removes moisture, the food shrinks and becomes lighter. When you are ready to use the food, you replace the water and the food returns to its normal size and shape.

The heat used in drying should never exceed 150 degrees. If higher temperatures are used, the food cooks. You should never try to speed up the drying process by raising the temperature.

Low humidity will speed up the drying process. Air movement will speed up the drying process by moving the moist air away from the surface of the food.

There are two ways to dry food: Indoors or outdoors drying. Indoor drying can be accomplished with an electric dehydrator, an oven, or with just air. Outdoor drying is on the vine, solar drier or just in the sun.

Use only foods that are fresh and in prime condition for drying.

An old friend of mine still remembers the method they used when he was a child in the South as follows:

Green beans were strung by using a needle and strong thread. They called these leather breeches. Tie a knot in one end of the thread and push the needle through the center of the beans, pushing the beans towards the knot. When you get 2 or 3 feet of beans on the string, hang the beans up by the end in a warm dry area, but out of direct sunlight. Let them hang until the beans are dry. Store them in a paper or cloth bag until ready for use.

Peas, when the peas are ripe, lay them in the sun to dry. After they are dry, wait for a windy day. Place them on a sheet and beat the hulls off with a stick. The wind will blow the chaff away and leave just the peas. Store the peas in a paper or cloth bag until use.

Corn, cut the corn off the cob, and lay in the sun until dry.

My friend says most vegetables can be dried without any special instructions if you just use your common sense. Use good sanitation practices. Do not attempt to dry vegetables which are badly bruised or rotten. Keep the vegetables clean and protect them from insects during the drying process. Window screens can be used to make a box allowing airflow but protecting the vegetables from flies, etc. Check the Reference Section for information on drying food and other methods such as pickling.

Another method of drying is to use your car. If you do not have gasoline to drive your car, you can use the car as a drier. Place your drying racks across the backs of the seats and just leave the windows slightly cracked. Sitting in the sun with the window just cracked, the air in the car can get very hot. Just remember to not let it get over 150 degrees. This speeds up your drying process and gives you some protection against insects and dirt.

Electric dehydrators are available in many discount stores, sporting goods suppliers and on the internet. They work well. A friend of mine buys frozen vegetables when they are on sale. He then runs them through his dehydrator and places them in plastic bags. The plastic bags are then packed in five gallon buckets. He has some that are several years old and are still fine. If you decide to do this, I recommend you use a container that is impermeable to oxygen and an oxygen absorber.

Cheese Waxing

A lot of information has been circulating on the internet about cheese waxing. This is a practice that has a long history of use. Any cheeses you obtain that are waxed at the factory and do not require refrigeration should be fine for storage. However, the practice of coating store bought cheese at home with cheese wax is being advocated by some web sites. This can be a very dangerous practice. The following statement from Dr. Nummer of Utah State University explains the problem:

"From time to time, dubious methods arise for preparing and storing various food items. Current information being circulated about the merits of dipping cheese in wax and placing it in storage for many years can be placed in this category. Consider the science.

Waxing cheese is a method to minimize mold growth on the surface of cheese. It cannot prevent growth or survival of many illness-causing bacteria. In fact, it may promote anaerobic (absence of oxygen) bacteria growth, such as botulism. The practice of waxing cheese for storage is considered extremely unsafe.

Before the days of refrigeration, cheese was dryer and fermented to a lower pH (higher acid). These types of cheeses were traditionally stored at room temperature with wax covers. The very low pH and fermentation byproducts could inhibit foodborne illness bacteria. An example is parmesan-style cheese. Acid, dryness, and fermentation byproducts make this cheese storable at room temperature.

Today, many cheeses are made strictly for storage under refrigeration. These cheeses may not have a low pH and other factors created in the manufacturing process to prevent illness-causing bacteria growth because the manufacturer knows the cheeses will be kept refrigerated. If someone waxes this cheese and places it in food storage, there is no science indicating any level of safety. In fact, there is evidence of the opposite. Placing cheese meant for refrigeration at room temperature is a significant risk and hazard for foodborne illness."
Brian A Nummer, Ph.D
Extension Food Safety Specialist
Director Food Safety Specialist
Utah State University

Brining

Brining is commonly used for the preservation of meat and fish. The basic process of brining is to add approximately 8 lbs of salt to 5 gallons of water. A method of determining the correct concentration is with a raw egg. The ideal brine has enough salt to float a raw egg. You will need enough brine to submerge the meat or produce without any portion being exposed to air. Some meat products might require being weighed down to stay submerged. Leave the food in the brine until ready to consume.

Use canning salt for brining. This has no additives. Most stores stock canning salt in their canning supply section. Using salt with additives or impurities can produce less than desirable results, especially with fish. Fish must be cleaned prior to brining.

Any food grade HDPE, PP, or polycarbonate container is appropriate for brining. These materials can withstand the salt in brines. These containers will normally have the recycling number two (see chart in Water Storage Chapter 3).

Generally, food storage containers sold at restaurant supply stores are made of food grade HDPE, PP, or polycarbonate. The interior of ice chests and freezers are made of food grade HDPE. Any white, opaque plastic bucket that contains food for human consumption is made of food grade HDPE.

Jerky - The jerky you purchase in stores is heavily treated with brines, spices, sugar and preservatives. It does not store well because it usually has too much moisture. Real jerky is just dried meat with maybe a little salt. Real jerky is made by cutting lean meat into strips about one eight to one quarter inches thick. This is then dried in the sun or over a low fire. Smoking with a hardwood (no pine or fir) will help preserve the meat.

Keep flies off the meat and store in dry bug proof containers.

Pemmican – is very nutritious and was widely used by the North American Indians. It is a mixture of pounded dry meat, dried berries and suet. Suet is the hard fat from around the kidneys of large animals. The dried pounded meat and the mashed dried berries are mixed with the melted suet. The mixture is then formed into balls and allowed to dry. If kept in a cool dry place, pemmican should keep for several months.

Chapter 7 – Recipes

This chapter contains recipes, substitutions, and cooking tips for foods that are commonly used for long-term storage.

Cooking and ways to save fuel

Whole wheat cereal - In a saucepan bring 1 part wheat to two parts water to a boil. Remove it from the heat and cover. Let it sit for at least 10 hours and it is ready to eat.

Wonder Oven Recipes - See Chapter 8 for instruction on how to make and use a Wonder Oven. Wonder ovens are inexpensive and a great way to save fuel. The following recipes have all been tested in a Wonder Oven and work well. With a little imagination you can improve on these and develop your own.

Rice

Two cups white rice
Three and half cups of water
Salt to taste

Put the rice into boiling water. Use a small pot that leaves little airspace. Place boiling pot in Wonder Oven and leave for approximately 40 minutes. Using the Wonder Oven requires less water than normal because there is no evaporation.

Macaroni, spaghetti, and noodles

Place pasta in a pot of boiling water, along with a little salt. Nest the pot in the Wonder Oven for 15 minutes. Any longer and they will be overcooked.

Meat Dishes

Prepare meat dishes like stew as normal and bring to a good boil. Nest the pot in the Wonder Oven and cover immediately with the top cushion. Make sure there are no gaps for the heat to escape. Remember, food cooks best if the pot is full. The food will continue to cook for 4-5 hours as long as you leave it alone.

When cooking large pieces of meat, hams, whole chickens, or roasts, the meat should be covered in liquid and boiled for 20 minutes. Put in Wonder Oven. After two to four hours, add vegetables and spices. Bring it to a boil and place back in Wonder Oven to finish cooking.

An option is to place the meat in a sealed baking bag. Place the bag in water and bring to a boil. Place pot in Wonder Oven and the meat will cook in its own juices.

Vegetables

Root vegetables or potatoes can be cooked by bringing them to a boil in a pan of water. Put them in the Wonder Oven. It takes about twice as long as normal to cook them in the Wonder Oven.

If you like waterless cooking, put your vegetables in a plastic bag before submerging in water and cooking.

In England during WW2, the following recipe was used in Wonder Ovens or, as they were known then, hay boxes.

Rabbits or roof hares (cats) could be caught wild and the English government encouraged people to keep them to provide food. This recipe uses rabbit and will work with a Wonder Oven, Dutch oven, or a conventional stove.

Ingredients:
One whole rabbit, cut into pieces
One tbsp vinegar
One oz flour
Salt & pepper
One-two oz drippings (lard or meat drippings)
Two bacon rashers, chopped (if available)
Two medium onions, sliced
Three medium carrots, sliced
One pint water or stock
One cooking apple
Fresh herbs (as available)

Equipment needed
Measuring cup
Tablespoon
Saucepan
Mixing spoon
Strainer
Knife

Chopping board
Peeler
Frying pan

Directions

Put the rabbit to soak in cold water with the vinegar for thirty minutes
Remove and dry well
Mix the flour with the salt and pepper and coat the rabbit pieces
Heat the drippings
Then add the rabbit pieces and cook steadily for about ten minutes or until golden brown in color
Remove from the pan
Add the bacon, onions, and carrots and cook for five minutes. Return the rabbit to the pan
Add the water or stock and the grated apple and stir as the liquid comes to a boil and thickens slightly
Add the herbs
Bring to a boil and simmer for thirty minutes
Quickly put dish into the wonder oven and leave for four to five hours
Serve with seasonal vegetables

Leavening agents

Wild yeast

One quart warm potato water
One half-yeast cake
One teaspoon salt
Two tablespoons sugar

Two cups white or whole-wheat flour

Mix the ingredients. Place mixture in a warm spot to rise until ready to use for baking. Keep a small amount of the yeast to use for a starter for the next batch. Between uses, keep the yeast starter in the refrigerator, or keep as cool as possible until a few hours prior to use.

Add the same ingredients with the exception of yeast to the wild yeast starter before your next baking. By keeping the yeast starter alive, yeast can be kept on hand indefinitely.

Sour dough starter

Two cups white or whole-wheat flour
Two cups warm water
Two teaspoons honey or sugar

Mix all the ingredients well. Place the mixture in an uncovered glass or crockery jar and place in a warm room. Allow the mixture to ferment for 5 days, stirring several times a day. Stirring will aerate the mixture an allow air to activate the yeast. Small bubbles will rise to the top and it give off a yeasty odor.

After each use, the starter will need to be fed. Replace the amount of starter you used with equal parts of flour and water. In twenty four hours, the yeast will be reformed and ready for use again. Store the unused portion in a glass or crockery jar. Keep the mixture in a refrigerator or the coolest possible place. Just prior to use, activate the mixture by adding two to three tablespoons of flour and the same amount of water. Sour dough starter can replace all or part of the commercial yeast in a recipe.

Old fashioned recipes for cooking with whole wheat and other storage foods

Whole wheat bread recipe

Six cups warm water
Two tablespoons yeast
One half cup honey
One Tablespoon salt
Eighteen cups of flour

Dissolve the yeast in 1 teaspoon of honey and a quarter cup of warm water. The water should feel about the temperature you would use for a baby's bottle. If it is too hot, it will probably kill the yeast. Wait five minutes; if the mixture foams and smells yeasty, the yeast is active. (This is called proofing the yeast. If you are using rapid rise yeast, you do not need to proof the yeast.) Add the rest of the honey, salt, and water.

Stir in twelve cups of flour. Pile four cups of flour on mixing board. Take about two cups of dough and place on pile of flour. Knead in just enough flour so that the dough is no longer sticky. Repeat this until all the flour is added. Put the remaining two cups of flour on the board and knead all the dough for ten minutes.

Place the dough in a greased bowl and set in an unheated oven with a bowl of water heated to about one hundred and forty degrees. Place the water under the bowl of dough and place a damp towel over the dough. With the help of the humidity, the dough should double in about forty five minutes. Divide the dough into greased pans. Let the dough rise until it doubles again. Bake at three hundred and fifty degrees for one hour.

Cracked wheat

One cup cracked wheat (coarsely ground)
Three cups water
Salt to taste

Mix cracked wheat and salt into boiling water and cook in double boiler for fifteen to twenty minutes. Pre-soaking the wheat can shorten the cooking time.

Hard tack

Four cups wheat flour
Water
Salt

Mix flour, salt, and water until the dough is moist throughout. Form the dough into 4 inch squares ½ inch thick. Place on greased cookie sheet. Take a fork and punch holes in the squares. Cook them until they are completely dry. There should be no moisture left in them.

Stored in paper or cloth bags and protected from insects, they should last a year or more. This is the famous hard tack used in the civil war. You will want to soak the squares prior to eating if you want to keep your teeth!

Sour dough biscuits

One and a half cups sifted flour
Two teaspoons baking powder
One quarter teaspoon baking soda (if starter is quite sour, use one half teaspoon soda)
One half teaspoon salt
One quarter cup soft butter
One cup starter

Sift the dry ingredients together. Mix in the butter. Then add the starter and mix well. Place the dough on a lightly floured board. Knead lightly until smooth and elastic. Roll dough one half inch thick; cut with a floured cutter. Place biscuits in a greased pan and brush with butter. Let the dough rise one hour in a warm place. Then bake in hot oven at four hundred and twenty five degrees for twenty minutes. Serve hot. This recipe makes about one dozen biscuits.

Sour Dough Hot Cakes

One cup starter
Two cups flour
Two cups milk
One teaspoon salt
Two teaspoons baking soda
Three tablespoons melted shortening
Two tablespoons sugar
Two eggs

About twelve hours before serving hot cakes, mix starter, flour, milk, and salt. Let stand in a cheesecloth covered bowl in a warm place. Immediately before cooking, add eggs, soda, shortening and sugar to the batter in the bowl. Mix well and bake on griddle. For thinner cakes, add more milk. The recipe makes about 30 cakes.

Cornmeal mush

One half cup cornmeal
Two and three quarters cups Water
Three quarters teaspoon Salt

Mix cornmeal into boiling water, stirring constantly. Add salt and cook for about half an hour. Serve with milk and sugar.

Cornmeal Jonny cakes

Combine one cup stone ground white cornmeal
One teaspoon sugar
One half teaspoon salt
One and three quarters cups milk

In a large bowl, combine cornmeal, sugar, and salt. Stir in three quarter cups of milk. This should make a thin batter. Drop batter by the tablespoon-full onto a hot well-oiled iron griddle and cook the cakes over low heat for four or five minutes until underside is browned. Turn the cakes over and cook them for five minutes more, or until brown. Serve the Jonny cakes very hot with butter and syrup.

Indian Fry Bread

Four cups flour
Three teaspoons baking powder
One and one half cups warm water
One half teaspoon salt
One cup powdered milk

Combine all the ingredients. Form the dough into flat rounds with your hands and fry in a small amount of oil.

Tortillas

One cup corn meal
One cup white flour
One half cup water
Three quarter teaspoon salt

Ground corn may be substituted for the flour and corn meal

Mix ingredients together and knead well. Let stand for 10 minutes. Form into the shape of a thin pancake. Add more flour or water if needed. Cook in an ungreased cast iron pan or Teflon coated skillet. Turn them over so that they do not burn.

Chili Beans

Two cups dried beans (red kidney or pinto)
Four cups water
One teaspoon salt
One teaspoon dry mustard
Two tablespoons sugar
One cup tomato sauce or catsup
One onion
One tablespoon chili powder

Soak beans overnight. Drain and add other ingredients. Cook for one half hour on top of stove. Turn heat down and cook until tender. An option is to place the pot in the Wonder Oven.

Woolton Pie – A garden recipe

This recipe became one of the most famous in England during World War II rationing. Lord Woolton, Minister for Food, promoted it as a healthy meal choice.

Ingredients for the pie:

One lb (450g) of dices potatoes
One lb (450g) cauliflower
One lb (450g) rutabagas
One lb (450g) carrots
Three or four spring onions or some sliced onion
One teaspoonful of vegetable extract or bouillon cube;
One tablespoon of oatmeal
Chopped parsley

Ingredients for the pastry:
This pastry uses only flour and lard (no sugar or egg) for six oz (170g) of pastry
Four oz (113.4g) of whole-wheat flour
Two oz (56.7g) lard
Equipment
Chopping board
Knife
Teaspoon
Tablespoon
Saucepan
Pie dish
Bowl
Rolling pin

Instructions for the pie:
Dice up the vegetables
Place in a saucepan with just enough water to cover.
Add the oatmeal and vegetable extract and stir.
Cook for ten to fifteen minutes.
Stir occasionally to prevent the mixture from sticking.
Remove any excess water and allow cooling.
Put the vegetables into a pie dish and sprinkle with chopped parsley.

Instructions for the pastry:
Place the flour into a bowl
Cut up the lard and rub into the flour with your fingertips until you have a breadcrumb like consistency
Add a tablespoon of water to make dough
Roll out very thinly
Cover the vegetables with the whole-wheat pastry

Bake in a moderate oven until the pastry is nicely browned and serve hot.

Substitutes

Egg substitute - Combine one teaspoon unflavored gelatin with three tablespoons cold water and stir until dissolved. Add two tablespoons boiling water and mix. This will substitute for one egg. Decrease the water in your recipes to allow for the extra water in the egg substitute.

One cup Sour cream: substitute one-cup fresh or powdered milk and one and a half tablespoon vinegar.

One cup sugar: substitute three quarter cup honey or one cup molasses.

Two cups tomato sauce: substitute three quarter cup tomato paste and one-cup water.

One-cup milk: substitute one half cup evaporated milk and one cup water or one cup reconstituted milk.

Condensed milk: substitute one half cup hot water, one cup powdered milk, and one cup sugar. Beat thoroughly.

Homemade baking powder

One half teaspoon cream of tartar
One quarter teaspoon cornstarch or arrowroot
One quarter teaspoon baking soda

Mix together. It makes one teaspoon baking powder.

Corn syrup

Mix one-cup sugar and two cups water. Cook until the mixture is thick. Substitute for corn syrup.

Pectin

Three pounds sliced, washed, tart, green apples (like Granny Smith) with peels and cores. Crabapples are the best. Small, green, immature apples of most varieties work, too.

Four cups water
Two tablespoons lemon juice

Wash, but do not peel, about seven large tart green apples. Cut them into pieces. Put them in a pot. Add four cups of water and two tablespoons of lemon juice. Boil the mixture until it reduces almost in half (about thirty to forty five minutes), and then strain it through cheesecloth or a jelly bag. Boil the juice for another twenty minutes. Pour it into sanitized jars and seal them to store in the refrigerator, freezer, or process in a water bath.

Cooking instructions for some storage foods

Dried apple slices one cup dry apples to one half cup water yields two cups fresh apples.

Beans- soften beans by adding three cups hot water and two teaspoons baking soda per cup of beans. Soak overnight, drain, rinse, and cook.

Dehydrated carrots- add one volume carrots to one volume water. Allow to set for twenty minutes.

White rice-add one cup rice to two cups boiling water and one teaspoon salt. Cook covered for about twenty minutes or until moist and tender.

Rolled oats- one cup of rolled oats to two cups of boiling water and add one quarter teaspoon salt. Stir and cook for two minutes.

Sprouts

It really does not matter how sprouts are utilized in food preparation, they will help you maintain good health and stamina. If you had enough sprouting seeds in your food storage, you could live a full year or more, eating only from your kitchen garden. Sprouts are the least expensive fresh vegetables you can procure and store!

It is virtually impossible for a family to store enough fresh vegetables to last a long period. By sprouting seeds, fresh vegetables are only two to three days away, year-round. Sprouts substitute for green vegetables and replace lettuce and other greens when they are expensive or unavailable. Get a variety of seed and learn to use them and you will have fresh green vegetables year-round, even if you cannot grow a garden. This makes sprouting seeds a high-priority item for your family's preparedness plan.

The amount of food value stored in such a small space is a bonus to a family's food storage program. Sprouting is a very easy way to increase the utility of many types of grains, seeds, and legumes or beans. Sprouts are very easy to prepare and utilize. Both equipment and supplies are easily found and readily available almost everywhere. It takes minimal effort to grow a crop of sprouts. Bringing sprouts to the table, ready to eat, takes less than ten minutes during the entire three day (average) growth period.

Compared to vegetables gardening, kitchen gardening with sprouts is easy. Sprouts require no fertilizer. In fact, all that is required is some water, air and a small nook where they can grow. Sprouts conserve energy. They require few resources during their sprouting cycle. You can eat sprouts raw, and any sprouted beans or gains cook much quicker.

Do not be afraid to try something new. There's not much you can do to hurt sprouts! After a few tries you will discover at which stage of sprout development, your family prefers different sprouted seeds. Some like sprouted seeds best after they have sprouted just forty eight hours, others when four to five days old when the sprout has more "chewiness" and has a more substantially developed flavor. Past this point, as the sprout is actually becoming a plant, they tend to become bitter and woody. Actually, sprouts can be used any time after the shoot emerges from the seed. But with some seeds, it is better to wait until the shoot is longer.

Basic Sprouting Equipment

Generally, the equipment for sprouting can easily be found in your home. Here is a list of equipment.

Quart Jar
Piece of cotton gauze, nylon net, or pantyhose top--any clean, durable fabric.
Strong rubber band (or sealing ring for quart jar);

Basic Sprouting method

There are a few basic rules for sprouting. Almost all seeds are done the same way. To utilize the Basic Sprouting Method, follow these general directions: Measure the right amount of beans, grain, or seeds for a batch, removing broken seeds and foreign objects. The idea is to grow your sprouts to fill the jar you are using. A quart jar will require a cup of wheat berries. The smaller the seed the smaller amount you have to use.

After a few tries you will find out the amount it takes to feed your family.

Place measured amount of seeds in a jar that is half-full of warm water. Cull out "floaters" or "sinkers" (when majority of seeds rest on bottom, cull the floaters — when the majority float, pick out the "sinkers")

Secure gauze (or nylon fabric) over the mouth of the jar with the rubber band (or jar ring)

Soak six through eight hours, in a warm location in the kitchen.

Then drain seeds well by turning bottle upside down. Leave it angled to one side in the sink or dish drainer for a few minutes. Rinse them again, in warm water and remove any objects. Allow to drain and place back in kitchen were you had it. (Be sure to place where it is warm.)

Drain and rinse seeds two to three times per day. Always drain well after each rinsing.

When sprouts attain the desired length, eat it all, seed, sprout, and roots -- This makes a healthy meal or snack.

It is best to eat the whole thing, but any leftover sprouts should be stored in the refrigerator to retard further growth. Sprouts generally achieve peak palatability, highest vitamin content, and potency within two to three days.

Suggestion for using sprouts

Following are a few ideas for using sprouts. You can eat them as fresh sprouts, in salads, teas, sandwiches, soups, gourmet entrees, casseroles, pancakes, or breads. You are limited only by your imagination.

Try some of the following ideas, before you go out and buy books on sprouting. Should you need more information, find a book by searching the Internet.

Soups - Corn, garbanzo, lentil, mung, pea, radish, or wheat sprouts can be used for flavor or thickening.

Rice - Add sprouts to rice dishes (barley, chia, pea, radish, watercress, or wheat) whole or chopped just before serving.

Steamed vegetables - Add whole bean, chia, clover, corn, garbanzo, lentil, mung, pea, radish, or wheat sprouts during the final few minutes of cooking.

Sautéed vegetables - Add beans, cabbage, corn, garbanzo, lentil, mung, pea, radish, or watercress sprouts. These highly flavored sprouts are especially good with sliced onion, a clove of garlic, and/or some green peppers.

Vegetable juices - Make juices from sprouts by mixing tomato juice, ground, barley, cabbage, clover, lettuce, radish, and wheat. Start with one type of sprout juice and add others until you like the taste.

Mashed potatoes - Try adding finely chopped sprouts to your potatoes.

Stir-fry - Add any sprouts for extra vegetables.

On July 9, 1999, the FDA issued a consumer advisory advising all consumers to be aware of the risks associated with eating any variety of raw sprouts, and advising persons at high risk of developing serious illnesses due to foodborne disease (children, the elderly, and persons with weakened immune systems) not to eat raw sprouts. The majority of these incidents involved alfalfa and unsanitary conditions.

Preparation of Ground Acorns

Many of the Native Americans used acorns as a staple in their diet. Acorns taste like a cross between hazelnuts and sunflower seeds and makes first-rate flatbread. In a long-term emergency, this is an excellent way to supplement your diet. Oak trees grow in almost every state and could provide large amounts of food.

You will find that there is a large variation in the amount of tannin in different species of acorn. I have found some in the Sacramento Valley that can be eaten without any leaching. They are just mildly bitter; others take a lot of leaching. Taste the acorns in your area and find the ones with the least tannin.

Gather several handfuls of acorns. All acorns in the United States are edible. Some contain more tannin than others do, but leaching will remove the tannin from all of them. Shell the acorns with a nutcracker, a hammer, or a rock. The meat should be yellowish, not black and dusty (insects).

Grind the shelled acorns. If you do not have a grinder, crush them, a few at a time on a hard boulder with a smaller stone, Indian style. Do this until all the acorns are crushed into a crumbly paste.

Leach (wash) them. Line a big colander with a dishtowel or other clean cloth and fill with the ground acorns. Slowly pour water through the colander, stirring with one hand, for about five minutes. A lot of brown liquid will come out. This is the tannin. When the water runs clear, stop and taste a little. If the meal is not bitter, you have washed it enough.

If you do not have a colander, tie the meal up in a clean cloth and soak it in buckets of clean drinking water until it passes the taste test.

Drain the mash, or squeeze out as much water as you can with your hands. One warning: do not let wet acorn meal lie about for hours or it will mold. Use the ground acorn mash right away because it turns dark when left out in the heat.

Acorn flour will not rise by itself. Indian breads were small, thin cakes baked before the fire on large, reflecting rocks. If you mix it with flour containing gluten in a ratio of one to one, it will rise. Acorn flour can be used as a thickener for soups and stews.

Uses for the brown acorn water

Save the brown water from the first leaching. The brown water should be kept as cool as possible. Over time, a mold may form on top of the water and you will need to boil the water again to kill the mold. The brown water may be used in any of the following ways:

Laundry Detergent: Two cups of the brown water can be used as laundry detergent for one load of clothes. Your clothes will smell very good but lighter colors (and whites) will take on a tan tint.

Traditional Herbal Home Remedies: The brown acorn water has medicinal properties.

Wash the skin with brown acorn water to ease the discomfort of skin rashes, burns, and small cuts. Externally it can be used to treat hemorrhoids.

Hide Tanning: Tannic acid or brown acorn water is used in the process of animal hide tanning. Soak the clean, scraped animal hides in the brown water. The name tanning comes from the term tannic acid, which was originally used to tan hides.

Indian acorn griddlecakes

Two cups acorn meal
One half teaspoon salt
Three quarters cup water

Combine the ingredients and beat to a stiff batter. Let stand for one hour. Heat one tablespoon of fat or oil in a frying pan. Drop batter into pan to form cakes about three to four inches across. Brown cakes slowly on both sides. These cakes can be stored for several days.

Chapter 8 - Cooking and Heating

Heating in an emergency - If you live in a cold climate, your best choice is a wood stove. A stove will provide heat, both for warmth and cooking. A propane or kerosene heater is your second choice; just make sure they are approved for indoor use. Keep in mind that it is easier to heat a small space.

In situations where people have been totally without heat in below freezing weather, they have survived by building a tent or shelter inside the house. In that small space, a kerosene lantern or a candle will provide a surprising amount of heat. If there are several people, your combined body heat will make a big difference. If you use a candle or lantern, beware of fire.

Matches - Everybody likes to think that they are Daniel Boone and that it is easy to start a fire with one match. It is just not that easy. Go out in your yard, try it, and when you think you are good at it, try it with wet wood. Matches are cheap; you can buy a lot for a few dollars. Store more than you think you will ever need. If nothing else, they will be good trading stock.

Most people do not know that some matches have a shelf life. Trinidad Match Limited lists the shelf life of their safety matches as 3 years. I spoke to a representative of Jarden Home Brands, the manufacturer of the Diamond Strike Anywhere Matches. Their representative stated that their matches do not have a shelf life, but if kept dry would last indefinitely. I suggest sealing them in a water tight container with a desiccant.

The Diamond Strike Anywhere Matches have a red tip with an extra chemical on the match, usually white in color that allows them to be struck and lit on any abrasive surface. These matches usually have a "bulls-eye" looking tip.

Strike-on-Box or safety matches require that the match be struck on the friction strip on the matchbox. Strike-on-Box matches do not have the extra chemical tip. Safety matches are useless if you lose or damage the box.

Wood stoves are a great way to go: they can provide heat as well as the ability to cook. Unfortunately, because of pollution in some areas of the country, the government is attempting to limit the days you may use them or ban them entirely.

Currently you can purchase many different styles of wood stoves. They are available as fireplace inserts or full old fashioned cook stoves. If you have the room, the old fashioned wood cook stoves are your best choice. They provide heat; cooking, and some even have a reservoir for hot water.

Depending on the efficiency of the stove and the climate you live in, you may need from one to five cords of wood per winter. Buy the best stove you can afford. An inefficient stove will burn a lot more wood and make you a lot more work.

Wood stoves come with a built-in problem. Make sure that they are correctly installed and vented. If not correctly installed, they can become a fire hazard. Follow the manufacturer's instructions and local building code recommendations for clearances from combustible materials. Having spent time in the fire service, I have witnessed many fires caused by incorrectly installed wood stoves. Just because yours has been in use for a while, do not assume it's safe. Sometimes it takes years before the installation fails and the fire occurs.

If you intend to depend on a wood stove, make sure that you stock the necessary implements for cutting and splitting firewood. This includes axes, splitting malls and an old-fashioned crosscut saw. Do not forget the files to sharpen them.

Solar ovens are one of my favorite solutions to the problem of cooking. In most areas of the United States, you can cook with a solar oven for a majority of the year. A friend in Wyoming has boiled water and cooked in his with the oven sitting on four feet of snow. You just have to place good insulation between the snow and the oven. You can bake bread, cook stew, and boil water in a solar oven. The only type you can fry in is a parabolic solar oven. A common question asked about solar ovens is will they brown food when you bake. The answer is yes.

Solar cooking is clean, keeps the heat outside, and the food tastes good. Obtain your solar oven and practice with it now. In an emergency, you do not want to waste food experimenting. A general rule for solar cooking is that it takes about twice the normal amount of cooking time. If you cut your food into smaller pieces, it will cook faster.

Once you put your food in the oven, try not to open the lid any more than necessary. The temperature can drop up to 100 degrees every time you open the lid. Do not use stainless steel or bright aluminum cookware - it will reflect the heat instead of holding it.

Quart glass jars painted black work well for cooking pots. Prior to painting a jar, run a piece of one-inch tape from top to bottom on one side. After painting, remove the tape. This will give you a window, so you can check on your food during cooking. Cast iron is great on partially cloudy days because it retains heat and keeps cooking temperatures even.

Other uses for your solar oven include pasteurizing water, (see WAPI in Chapter 3) killing insect infestations in grains or dried foods, drying firewood or tinder, and heating water for dishwashing or sanitation.

Many sites on the internet sell solar ovens in the $100 to 300 dollar range. Solar ovens are easy to improvise and can be made for under $30. A simple solar oven plan is shown in Chapter 15, and there are references to some excellent websites in the Reference Section.

The above solar oven works well. It is made of two cardboard boxes, one inside the other with newspapers shoved between them for insulation and lined with an old windshield sun reflector. The top is two pieces of tempered glass a neighbor gave me. The tire is just to keep it off the damp ground.

This stove will boil water and reach a temperature of about 240 degrees.

The solar oven in the above picture is a metal box lined with silver colored Styrofoam insulation. The lid is lined with aluminum foil. A piece of 1/8 inch Plexiglas is used for the top. Notice the thermometer on top of the pot.

I have cooked beans successfully with this stove. It will reach temperatures if about 250 degrees.

Rocket stoves - A simple homemade stove that can be made from old tin cans and scrap metal. This type of stove is extremely efficient; it will generate an amazing amount of heat burning small twigs and sagebrush. If you were faced with a shortage of fuel, this would be a great stove to own. It will burn the twigs, dried brush, and other small debris that most people ignore.

The stove shown below is made from a five-gallon can and some scrap 4-inch pipe, and works well. A plan for manufacturing one is in Chapter 15.

Rocket Stove

Camping stoves - In this class, I am including Coleman style camp stoves and others that run on white gas (Coleman Fuel), propane, or butane. The small backpacking stove will be discussed later in chapter 13.

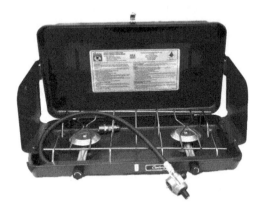

Coleman Propane Stove

The older Coleman stoves are a good product and are quite efficient. The problem with them is the type and quantity of fuel you have to store. White gas and Coleman fuel are highly flammable; storing any quantity is dangerous without adequate safeguards, and in many areas is a violation of fire codes.

Many of the new Coleman-style stoves run on small non-refillable propane tanks. Conversion kits are available so that these can be connected to larger five-gallon propane tanks. This lets you cook for longer periods of time and reduces your cost. Propane is much safer to store than white gas.

If your home runs on propane, you may have several hundred gallons available in the tank in your yard. When the power fails, most modern propane appliances will not run without electricity. Make sure you have the necessary plumbing supplies on hand to connect your stove to your propane line. Any good hardware store should be able to supply the needed parts.

If you go to one of the better camping/backpacking stores, you will find a selection of excellent small backpacking stoves. Most of these are so small that they are not practical for everyday use. They belong in your 72-hour kit. I would not depend on them for my main cook stove. They utilize bottled butane, propane, wood and flammable liquids for fuel. My personnel choice would be one that uses wood or flammable liquid. You have a better chance of finding replacement fuel in an emergency. They are light, easy to use, and quite practical for a few days use.

Kerosene stoves are commonly used in many third world, African and Asian countries. The brand that I am familiar with is the Butterfly stove from Malaysia. It is inexpensive, reliable, and available. There are several sources on the Internet. In the Reference Section, I have listed a company that I have dealt with and found reliable. The Butterfly stove is available in one and three burner models. The single burner model will run for approximately 12 hours on a half gallon of fuel. The wicks are easily changed and are good for up to six months.

While the storage of fuel is always a bit of a problem, I feel these stoves are very worthwhile. Ten gallons of kerosene would provide you with approximately 240 hours of cooking. Depending on your climate, if you share the cooking with a solar oven on sunny days, you could cook for a year with a relatively small amount of kerosene (10 gallons).

Butterfly oven

Butterfly also manufactures a light weight portable oven that works as a companion to the Butterfly stoves. It is capable of baking bread. The oven sits on top of the stove covering one of the burners. It will work on either the one or the three burner models. It is open at the bottom and can also be set on a grill or over hot coals. If you set it over hot coals, you need to use flat rocks so that it is not in contact with the coals.

Single-burner Butterfly stove-The glass bottle holds the kerosene and feeds the stove by gravity.

Sixteen wick Stove

Butterfly stoves has come out recently with a new version known as the 16 wick stove. This stove produces more heat than previous models and works well with the oven for baking bread. The downside to this stove is that it uses about twenty-five percent more fuel.

In an emergency, a Butterfly stove can serve as a heater. Place a firebrick, large rock or a piece of heavy metal on the burner, and turn the heat up. The mass will work as a heat sink and help disperse the heat evenly. This will not be as efficient as a kerosene heater, but it will warm a small room.

When first lit, all kerosene stoves will smoke a little since the burners do not provide maximum efficiency until it they are warm. Cracking a few windows will solve the problem. When you are through using a kerosene stove, do not blow it out. Turn down the wick and let it burn out. This will use up the kerosene vapors remaining in the wicks, and help eliminate kerosene odors.

Butane stoves

There are small one-burner butane stoves being sold in many retail stores. My experience with the stoves is that they work well. The downside is that the fuel comes in aerosol cans. If you intend to utilize one of these stoves for any period of time, you will need to store a large number of fuel cans. While the stoves are inexpensive, the costs of the fuel canisters mount up quickly.

Butane Camp Stove

How to Cook Your Food over an Open Campfire

If you are like most of the population of the United States, you have not lit a campfire in many years. Lighting campfires takes practice and experience. In an emergency, matches will be in short supply. Learn how to start a fire now. I have seen inexperienced people use whole boxes of kitchen matches and still not have a fire started. Learning to light fires cannot be stressed strongly enough. If you cannot start a fire in the rain with wet wood, you have not practiced enough.

The pioneers used tripods to hang their cooking pots over open fires. These are easy and cheap to duplicate with ½-inch rebar and a short length of chain. These items are available at any building supply. Remember to be fire safe.

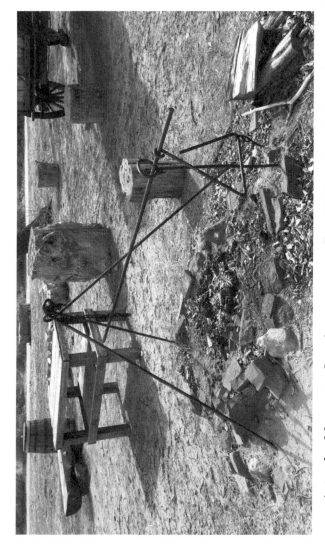

A typical layout of a pioneer campfire. Notice the chain to hang pots from, it is doing double duty by holding one end of the spit

Dutch ovens - These were widely used by the pioneers due to their versatility. Dutch oven cooking is an art all by itself. Before purchasing a Dutch oven, consider the following

- Buy only Dutch ovens with legs. There should be three legs, firmly attached.

- The lid should fit tight with no gaps around the rim and have a lip around the top to hold hot coals.

- The casting and thickness of the metal should be even, especially around the rim. Large variations will create hot and cold spots during cooking.

- Make sure the lid has a loop handle tightly attached to its center.

- The bail or wire handle should be attached firmly to the pot. The bail should be easily movable and strong enough to carry or support a heavy pot full of stew.

- Have a good pair of leather gloves, and something to use to lift the lid. There are manufactured lid lifters available through most of the better hunting or camping suppliers.

Dutch ovens can be used over an open fire. They also work well in a solar oven or with practically any other type of stove. They can be hung over an open fire or placed in the hot coals. You can then scoop up hot coals and place them on the lid. Put 2/3 of the coals on the top and 1/3 under the bottom. Periodically turn the lid ¼ turn clockwise and turn the bottom ¼ turn counter clockwise. This prevents hot spots. This gives you even heat on the top and bottom, which permits you to bake bread, rolls, and even cake.

If you have the older style Dutch ovens the bottom side of the lid is flat. After cooking your meal, you can take the still hot lid off and place it upside down in the ashes. Keep it flat and you can cook old-fashioned hoecakes, which are made with a corn meal batter, and then cooked like pancakes.

Many of the Dutch Ovens that you currently see on the market are aluminum. I would recommend that you avoid them and stay with cast iron. Aluminum is lighter in weight, but has a tendency to cool rapidly. With aluminum, it is much harder to keep an even heat while cooking. This makes it harder to bake.

By placing charcoal briquettes on the top and bottom, you can regulate your cooking temperatures. The following chart shows the number of briquettes required for specific cooking temperatures.

Temperature	8" Oven		10" Oven		12" Oven		14" Oven		16" Oven	
Degrees F	Top	Bottom	Top	Bottom	Top	Bottom	Top	Bottom	Top	Bottom
300	10	4	12	6	14	8	16	10	18	12
325	11	5	13	7	15	9	17	11	19	13
350	12	6	14	8	16	10	18	12	20	14
375	13	7	15	9	17	11	19	13	21	15

Note: Adding one set of briquettes (one on top and one on bottom) will raise the temperature of the Dutch Oven approximately 25 degrees. Or conversely, removing one set of briquettes will lower the temperature by 25 degrees.

A typ[c]al Dutch oven camp kitchen. Notice the coals on top of the lids. This provides an even heat.

Many new cast iron frying pans and Dutch ovens come from the manufacturer with a waxy coating. This needs to be burned off prior to use. The best way is to turn the oven upside down over an open fire or charcoal. When they heat up, you will see the waxy coating bubble up. Let this burn off and wipe with a clean rag. You are now ready to season your pot.

To season a cast iron pot, coat the pot with lard, bacon grease, or Crisco and bake at 250 degrees for 4 hours. Do not use a liquid vegetable oil or the pan will be sticky and not properly seasoned.

After cooking, wash the pan while still warm in hot water and scrape the pan if needed. Do not use scouring pads or soap; they will break down the pan's seasoning. If your pan rusts, it needs to be re-seasoned.

Hobo stove - Excellent for use in cold weather.

Wonder ovens - are insulated cushions placed in a box. You place a hot boiling pan of food in the center and let it finish cooking under its own heat.

Cooking was very challenging during World War II in England. Frequent air raids could make it difficult to cook a hot meal. Once the sirens sounded, everyone would head for their bomb shelters to wait it out. No one wanted food to spoil because rations were not replaced if anything happened to them.

In the centuries before this, farm workers had sometimes used a hay box (early version of wonder oven) to keep food slowly cooking. This method regained popularity. A stew or soup could be left, packed tightly in hay inside a box. It would slowly cook until needed. This would provide a hot meal for a family when the raid was over.

You will need a little less than three yards of cotton cloth to make a Wonder oven. Cut out four pieces for the lid and four pieces for the bottom using the pattern on the following page:

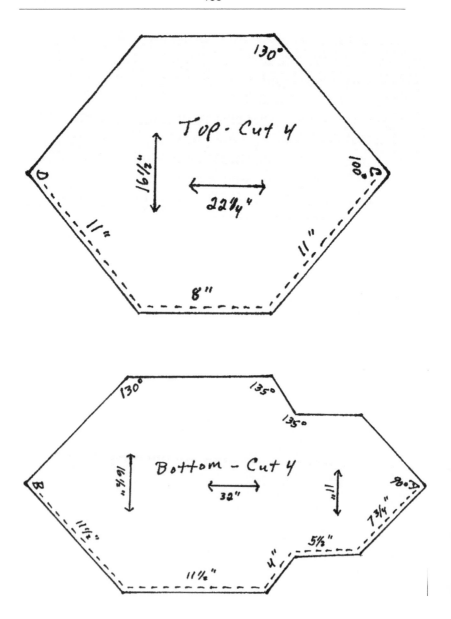

To make the bottom, sew two of the four pieces together along the longest side. This is from point A to point B on the diagram. Now sew the other two pieces together the same way. You now have two larger pieces. Open the two pieces and place one on top of the other.

Sew all the way around leaving a six-inch opening so that the material can be turned inside out. This opening is then used to fill the cushion with either styrofoam or wood shavings and then sew the opening shut. It takes six gallons of styrofoam. This leaves you with a funny shaped cushion. Now push the narrow part of the bottom into the larger part to make a nest for a pot to fit in.

Making the top is similar to the bottom. Take two of the pieces and sew them together from point C to point D on one side. Now sew the other two pieces together the same way. Place the two pieces on top of each other and sew all around the edge. Leave a six-inch opening so that you can turn the material inside out and fill the cushion with four gallons of Styrofoam. Sew the opening shut.

Place the bottom in an approximately twenty-quart container. It can be plastic. Find a pot with a tight fitting lid that will nest in the bottom. Place the top on and you are ready to cook. When you place the hot pan in make sure the pan is completely enclosed by the Wonder Oven. If there is any place for the heat to escape the oven will not work.

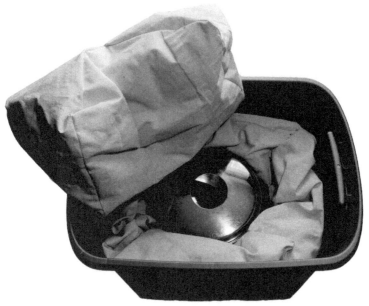

Wonder oven with cooking pot in place, ready for cooking in 20 quart plastic bin. Just put the top cushion into place.

Wonder oven with top in place covering pot.

Boil your food on the stove first for a few minutes until the food is heated all the way through. Use a pot that fits the food; do not leave a large air space in the pot. As you bring the food to a boil put the lid on so that it gets hot. Rapidly place the pot in the nest made by the bottom of the oven. Quickly cover with the top, making sure there are no gaps for heat to escape.

When cooking with a Wonder Oven, remember that the fuller the pot, the better it will cook. When cooking something like a chicken, make sure the liquid covers the contents and let it boil for 15 minutes before it goes in the Wonder Oven.

The cushions in the Wonder Ovens are washable if filled with styrofoam.

Do not leave foods in the Wonder Ovens for extended periods of time after they are cooked. Foods that are allowed to sit in there until they are cool can develop bacteria and cause food poisoning.

Cooking times:

Food:	Boil time:	Wonderbox time:
Rice	5 min	1-1.5 hours
Potatoes/carrots	5 min	1-2 hours
Soup and stew	10 min	2-3 hours
Bread	12 min	3 hours
Pintos or other beans with meat	15 min	5 hours
Split Pea Soup	10 min	3-4 hours
Muffins	10 min	4-5 hours
Oatmeal	1-2 min	1 h or hold overnight
Meat loaf	15 min	4 hours
Winter Squash	5 min	1-2 hours
Steamed bread	10 min	2 hours
Chicken	6 min	2-3 hours
Beef Roast	15 min	5-6 hours

Approximate cooking times for the Wonder Oven.

If your food in not finished, you can bring the pot back to a boil and put it back in for further cooking.

Just remember

The Wonder oven conserves fuel.
Food cannot be burned or dried out.
The Wonder Oven can be used both indoors and outdoors.
It can be used to keep things cold in place of a cooler.

Chapter 9 – Lights

The premise of this chapter is that the electric grid is nonfunctioning or unreliable for extended periods. You are dependent on your own resources.

Days get a lot shorter without lights. Once the sun goes down, there are not a lot of things you can do. If you read by candle or a poor flashlight, you will end up with a headache. Crime also increases after dark. Living without lights can become very depressing.

Groups who have experienced this say that having two or three LED headlamps made a big difference. For the simple things like washing dishes, preparing food or doing stuff around the house, they are worth their weight in gold. Get LED headlamps that use rechargeable batteries.

Generators are handy for short-term power outages. It is not possible for most of us to store enough gasoline to make generators practical for long-term use. A good 5000-watt quality generator can cost upward of $2700. For $500, you can get a cheap Chinese made 5000-watt generator. The cheap ones will be unreliable, noisy, and use more fuel.

There are generators available that run on gasoline, diesel and propane. Diesel has a longer storage life than gasoline. Propane's storage life is almost indefinite. See Chapter 10 for information on storing fuels. Diesel and propane generators are more expensive but have advantages for long term use.

Unless you have the capability to store several hundred gallons of gasoline, diesel or propane, I do not recommend that you invest your money in a generator as a solution to long-term power outages. This money can purchase a lot of food and other supplies.

If you live in an area with occasional power outages from storms two to three times a year, a generator may be a good investment. With a good generator, 5 – 10 gallons of gas will get you by for a few days. To determine how much fuel to store, keep in mind that generator manufacturers rate their fuel consumption on running with a load at ½ capacity.

The typical home generator is fueled by gasoline and produces between 1000 to 5000 watts. If you want to run more than your TV and a few lights, you will need at least 5000 watts of power.

Most people buy a generator, start it once, and put it away in the garage. Since most generators run infrequently it is best to place them in mothballs.

First, drain the fuel from the gas tank. Let the generator run until the carburetor runs dry. The motor oil should be left in the engine. If it sits without running for several years, the oil would need to be changed. Check your oil levels and if the oil is black, it may be time to change it. Keep your generator dry and out of the sun.

When it is time to start your generator, you will have to pull on the starter rope a few times to get the gasoline into the carburetor. After the generator is running, plug in your lights or appliances.

Know the starting load that your appliances or lights will draw so that you do not overload your generator. Use heavy-duty extension cords to reduce the resistance. Light duty cords can heat up under heavy loads and start fires.

Heating systems and Major appliances		Household appliances	
Electric heat	4,000-6,000	Fry pan	1,160
Pellet stove	600-1,000	Iron	1,100
Gas and oil heat fan	500	Toaster	1,100
Heating systems and Major appliances		Lawn mower	1,000
		Coffee maker	850
Range	12,000	Hair dryer	700
Hot water heater	4,500	Vacuum cleaner	700
Clothes dryer	4,350	Blender	290
Space heater	1,300	Blanket	170
Dishwasher	1,190	Stereo	160
Microwave	650	Sewing machine	100
Refrigerator	425	Radio	75
Washing machine	375	Crock pot	70
Color TV	115	Light bulb (60W)	60
Black and white TV	75	Clock	2

The above table shows some typical appliance loads

Warning- On occasion generators have been hooked up to back feed into the house wiring. **This is not recommended**. If you do this and fail to isolate your house wiring from the electrical grid correctly, you are feeding power back into the electric grid and can electrocute a lineman working to repair the system. If you wish to use your house wiring with a generator, get a certified electrician to install a transfer switch that complies with your local building codes.

Warning- Gasoline, diesel and propane engines give off carbon monoxide. Do not run generators in your house or garage as the fumes can kill you.

Remember running a generator makes a lot of noise and everyone in your neighborhood will know you have one. Quiet generators are expensive, loud is cheap.

DC to AC power Inverters - A "DC to AC" power inverter electronically converts DC power from a car battery or other DC power source to 60 hertz AC power at 120 volts like in your home. Most small inverters have standard household AC grounded power outlets. They can be used to power small appliances like radios and lights.

- Most small inverters require 12-volt input.

- The 150-Watt models can plug directly into a car cigarette lighter.

- Larger units in the 300-Watt range often come with both cigarette lighter cable and alligator clip cables for connection direct to a battery. The cigarette lighter cable is for use up to about 150 Watts. If you need more power than that, you must use the other cables that are included to connect directly to the battery.

- The 600 and 1500-watt models may be hard wired directly to a battery or other DC power source, or attached with included jumper cables.

- The larger 3000-Watt models must be wired directly to a battery or other DC power source.

Please note: A typical cigarette lighter socket in a car has a 10-15 amp fuse, therefore the most power you will be able to use through a cigarette lighter socket will be in the 120-180 watt range. Power demands above that range must be via direct connection to the battery.

Candles - While this is one of the oldest methods of lighting, it is not the best. Even good quality wax candles give off significantly less light than a kerosene lantern.

A ¾ inch candle should burn approximately 30 minutes per inch of length. By increasing the candle diameter to 1 inch you will double the burn time to 1 hour per inch.

The biggest advantage of candles is that if you shop carefully, you can purchase them quite inexpensively. After Christmas and other holidays, you often see them on sale. You can buy a few every month and gradually build up your stock. Some of the 100-hour candles sold by survival supply stores contain liquid paraffin fuel and are not a true candle. The flame they provide is a little brighter.

Remember, with all candles you are dealing with an open flame and they are a fire hazard. Stock some good candleholders. You can improvise a candleholder by sticking one or more candles into a plant pot filled with dirt and placing a piece of shiny metal, aluminum foil or a mirror behind it.

Butane and propane lanterns - These are good reliable **products.** Most provide plenty of light, and use fuel efficiently. Some of the propane models can be converted to work with 5 gallon or larger propane tanks. If you intend to use this type of lighting, remember to purchase at least a dozen extra mantles and any fittings you need to convert to larger propane tanks. Read the instructions that come with them. Butane lanterns cannot be converted to propane. Avoid cheap foreign manufactured knock-offs.

Solar lighting -A friend of mine has created a simple 12-volt solar system for a few hundred dollars. He mounted two small solar panels on the south side of his home. He uses them to keep a pair of 12-volt deep cycle marine batteries charged. This provides power to several 12-volt lights located throughout his residence. Check the Reference Section under solar for sources.

Flashlights - A good flashlight can save your life. You need them in your home, car and 72-hour kit. With the advent of the LED bulb, flashlights have become smaller and more efficient. Always keep your flashlights in the same spot so during an emergency you can find them in the dark.

There are numerous flashlights on the market that do not require batteries. They operate by shaking, squeezing or cranking. Before you buy any of these, check to see where they are manufactured. Avoid any made in China. Many of the ones on the market are of poor quality and are not reliable in an emergency. I have not found any of the flashlights that function by shaking to meet my requirements. The only thing I can say about them is that they are better than nothing.

Last year I purchased some solar powered flashlights manufactured by HYBRID SOLAR LITE. These operate by both battery and solar power. I have tested them on only solar power (batteries removed) and they work well.

Consider getting an LED headlamp. They let you have both hands free while working. Some of the new ones will provide 40 hours of light with three AAA batteries.

LED bulbs have extremely long lives, but it is not practical to replace bulbs in these flashlights when they go bad. Good quality non-LED flashlights like Maglights have spare bulbs stored inside the base. If your flashlight uses incandescent bulbs, be sure to store extra bulbs.

Extra batteries should be kept refrigerated or as cool as possible. This prolongs their shelf life. There are currently small solar chargers on the market that will work with rechargeable batteries. They are available on the internet and many of the larger chain electronic stores carry them. I would recommend getting a solar charger and rechargeable batteries as well as a good stock of standard batteries. If possible, standardize the size of batteries and type of flashlight you use. You can always cannibalize one to repair another.

Generic alkaline batteries are the most common non-rechargeable battery you will encounter. These cells offer two to four years shelf life according to their expiration dates. I have been able to store alkaline batteries for much longer than their recommended shelf life by keeping them cold. Tests on alkaline batteries show that if properly stored, they still provide a functional capacity of 75-80% after 10 years. Store your batteries in separate small packs, so that if one fails and leaks, damage is limited.

Nickel cadmium (NiCd) and nickel metal hydride (NiMH) batteries. Nickel-metal hydride (NiMH) batteries are the newest, and most similar, competitor to NiCd batteries. Compared to NiCd, NiMH batteries have a higher capacity and are less toxic, and are now more cost effective. However, a NiCd battery has a lower self-discharge rate (for example, 20% per month for a NiCd, versus 30% per month for a traditional NiMH under identical conditions), although low self-discharge NiMH batteries are now available, which have substantially lower self-discharge than either NiCd or traditional NiMH

Low self-discharge NiMH batteries are often sold as "pre-charged" or long shelf life NiMH batteries. The most common are manufactured by Duracell or Eneloop. Sanyo makes Eneloop batteries. Sanyo claims that they are rated for 1500 recharges, and if left charged, will have 75% of their charge left after three years.

Non-rechargeable lithium batteries are the best for long-term storage. Their storage life is listed at 10 years, but it is probably longer. They are excellent at low temperatures. They are available in double and triple A as well as CR123. They are quite expensive and not as available as alkaline batteries

Battery technology is advancing rapidly. There are many new batteries under development. Watch for new and improved types.

Always test dead batteries prior to throwing them away. Just the other day I had a flashlight go dead. Testing the batteries revealed that only one of the batteries was dead. The other was still working. I keep a little inexpensive tester that I purchased at an auto parts store for a couple of dollars. It has saved me more than it cost in salvaged batteries.

Kerosene lanterns were in common use prior to the advent of electricity. They provide good light and are efficient and simple to use. You can purchase hurricane lamps quite inexpensively for under $10.

Most kerosene lanterns share a few things in common. The majority of kerosene lanterns, other than the Aladdin, have a flat wick. The width of the wick determines the size of the flame. The larger the flame, the more light it produces. The most common wicks run in 1/8 increments from ½ inch to 1 inch.

Kerosene lanterns are made with either a short or tall chimney or globe. Both have advantages. The short ones are more compact and easier to store. They are also easier to clean. The tall ones create a better draft, which results in a bit more light.

How to get the most light from your lantern - Make sure the globe or chimney is clean. Even with the lamp burning at its greatest efficiency, it produces soot. A thin veil of soot will build up over the inside of the globe without being noticed. Wipe the chimney clean every time you refill the fuel.

Keep your wick trimmed. The very end of the wick will become charred with normal use. This affects how the fuel is drawn up into the wick and burned. Trimming the wick periodically will keep your light bright. Normally you will have to trim a one eight to one quarter inch off the burned end of the wick. Be very careful to trim the wick straight across. A nice square cut will provided the brightest light.

The type of fuel you use will make a big difference. In the U.S. there are several different fuels. 99% paraffin, kerosene, and synthetic kerosene are some of the fuels sold under the name lamp oil. Synthetic kerosene burns as brightly as kerosene, but with fewer odors. Kerosene is bright but has a strong odor. Paraffin has the least odor, but does not burn as bright.

Lighting and adjusting the flame correctly. Adjust the unlit wick so that it protrudes just above the wick tube in the burner. Raise the chimney and light the wick, then lower the chimney. As soon as the whole top of the wick is burning, raise the flame height ¼ to ½ inch. Let the lantern warm up before adjusting the flame further. Once the lantern is warm, it burns better. Adjust the flame to its brightest level. If the flame is too high, the chimney will rapidly begin to blacken with soot.

Kerosene lanterns provide approximately 68 lumens of light or about 6 candlepower. Older Petromax kerosene lanterns provide 1300 lumens of light. They are much more efficient than a hurricane lamp. The following table compares the fuel use to the amount of light given. You can see that a Petromax lantern is comparable to a 100-watt light bulb. The downside to the Petromax is that they are more expensive, costing in the area of $130.00. If you decide to obtain a Petromax, be sure to get a real one, not one of the cheap copies made in Asia.

Fuel consumption

Light source (fuel)	Light output lumens (lm)	Fuel consumption	Efficacy (lm/W)
100 W bulb (electricity)	1340	100 W	13.4
Hurricane (kerosene)	68	16 g/hr	0.35
Petromax (kerosene)	1300	80-90 g/hr	1.27
Fluorescent tube, 40 W (electricity)	2400	40 W	60.00

Information in this table is from the Nimbkar Agricultural Research Institute (NARI), Phaltan-415523, Maharashtra (India).

Aladdin Lamps- Aladdin lamps are a kerosene lamp that uses a round wick. They are more expensive, but provide a lot more light. They are bright enough to read by without causing eyestrain. Buy at least twice as many mantles as you think you will need, as they are quite fragile.

BriteLyt lanterns are a new product manufactured by Petromax. They claim to run on a variety of fuels. Kerosene, alcohol-based fuels (with adapter), mineral spirits, citronella oil, gasoline, Biodiesel, diesel fuel,

Coleman fuel, JP fuels, and almost every flammable fuel available on the market. The company is currently manufacturing them for the U.S. Military. They have two different size lanterns, the CCP 150 & CCP 500. The CCP 500 can be fitted with attachments for heating and cooking.

The CCP 150 can provide a light equal to a 300-watt bulb and the CCP 500 equal to a 550-watt bulb. They have a reputation for having a bit of a learning curve and come with an instructional DVD. Watch the DVD and practice lighting your BriteLyt while times are good.

From left to right: battery operated LED lantern, BriteLyt CCP 150, Propane lantern, Coleman Lantern, Hurricane Lantern, Kerosene Lantern.

Rollable Solar Panels – There are currently several companies that produce rollable solar panels that can provide you with 12 volts of power. These are available in various wattages running from 5 to 28 watts. They are small and light enough to carry on your backpack.

Rollable solar panels can be used to charge batteries for lighting systems, car and boat batteries, some electric coolers, cell phones and computers. The panels I have tried are by PowerFilm and are built to be as versatile as possible and easy to incorporate into homegrown systems.

A charge controller is recommended where the load on the battery is less than or approximately equal to the power produced by the solar panel to prevent overcharging of the battery and possible damage to the battery.

There are numerous accessories available including chargers for Lithium, NiMH, NiCd, Sealed Lead Acid, and Flooded Lead Acid batteries. Lanterns, fans and various adapters for computers, cell phones, and cars are readily available.

In addition, fold up solar panels are available in sizes up to 60 watts. These can be connected together to provide more power.

Chapter 10 - Fuels

Warning: Cooking, heating, and lighting units (this includes gasoline motors) that burn fuel of any type give off gases. Most unvented appliances should not be used in a closed area because they give off carbon monoxide. Carbon monoxide will kill you if not properly vented.

Alcohol burns clean and gives off almost no products of combustion. Kerosene gives off carbon dioxide, which is less dangerous than carbon monoxide. Alcohol and kerosene are the safest of the fuels to burn in unvented appliances. Alcohol and kerosene lamps can be used in your home.

None of these appliances should be left on overnight or while sleeping. Do not under any circumstance burn charcoal in an enclosed area as it gives off carbon monoxide and will kill you.

Carbon monoxide is not toxic in the normal sense. It combines with the hemoglobin in your blood and starves the cells of needed oxygen, which results in internal suffocation. Breathing a concentration of 1000 parts per million will kill you in a relatively short period of time. Carbon monoxide is a colorless, odorless gas.

The symptoms of carbon monoxide poisoning can include headache, dizziness, and confusion, pink to reddish skin color, nausea, vomiting and fainting. The only treatment is to give victims oxygen or exposure to fresh air.

I am sure some of you remember your grandparents violating some of these rules in past years. You also need to remember that their houses were often drafty and not well insulated. Modern homes are much better sealed.

Solar is one possible solution to the fuel storage problem. If you are dedicated enough and willing to spend the money, you can go completely off the grid. I am not going to attempt to recreate the wheel by telling you how to put in a full solar system. Refer to the Reference Section for further information.

Simple solar systems can be improvised for a couple of hundred dollars using several 10 - 15 watt solar panels, one or two 12 volt sealed lead acid 7.5 amp hour batteries (burglar or fire alarm system batteries) and some LED blubs. Simply use the panels to charge the batteries during the day and run the LED off the batteries at night.

Charcoal is a dark grey residue consisting of impure carbon obtained by removing water and other volatile constituents from animal and vegetation substances. Charcoal is usually produced by slow pyrolysis, the heating of wood or other substances in the absence of oxygen.

Charcoal is currently readily available, and relatively inexpensive. Cooking with charcoal is easy most of us already have a barbeque. In an emergency, you may want to conserve charcoal by using the smallest practical barbeque. Dutch ovens work very well with charcoal. You can regulate the temperature by the number of briquettes you use. See Dutch oven cooking in Chapter 8 to learn how to control the temperature.

In many third world countries, charcoal is used for everyday cooking by a large percentage of the population. If used indoors, charcoal can cause serious health problems. Reports indicate that acute respiratory infection from indoor cooking fires is the number one cause of death in children under five in third world countries. In addition, carbon monoxide (CO) is a combustion product and can kill you without adequate ventilation.

Besides cooking, charcoal has other uses such as improvised water filtration systems, blacksmithing, and some medical use.

Making charcoal is surprisingly easy. You can make small quantities in your backyard. The method I describe below is the retort or indirect method.

Set a 55-gallon barrel on a metal stand. Cut an opening in one end of the barrel. Build a door that closes tightly. Install a 2-3 inch vent pipe out the back.

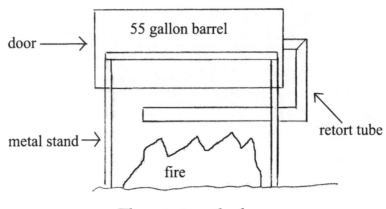

The retort method.

Route the vent pipe underneath as shown in the above diagram. Plug the end of the vent pipe. Drill holes in the top of the lower section of the vent pipe to create a burner. Place wood inside the barrel and shut the door. Start a fire under the barrel. When the wood in the barrel starts cooking, the volatile vapors will vent out through the pipe and burn. When the vapors from the vent pipe are no longer burning the charcoal should be done. This should take about three hours and can make up to 60 lbs of charcoal.

An older charcoal making method involved digging a pit and filling it with wood. The wood is then covered with straw and dirt. You leave vent holes at the top and bottom. Set the wood on fire. When the wood is burning well plug the holes at the top. The wood will continue to burn slowly creating charcoal. It may take a couple of days.

Always use clean dry wood. Never use pressure treated wood. The vapors will be harmful to your health.

Charcoal should be stored in a dry location; when wet it has been report to spontaneously combust.

During the early days of the automotive age and during World War II, cars were run on the volatile vapors given off by heating wood. This was a retort very similar to the one I have suggested for making charcoal. The big difference was it was attached to the vehicle and was portable. The vapors from the wood were vented into the carburetor. Charcoal was a byproduct.

Coal - One quarter of the world's coal reserves are found within the United States, and the energy content of the nation's coal resources exceed that of the entire world's known recoverable oil.

The American Indians were using coal when the first settlers arrived. Coal was plentiful and easily found on the surface. Today in parts of the country, coal is still widely used and readily available. However, it is unlikely you will find it on the surface.

If coal is available in your area, it is a good cooking and heating fuel. Be sure and keep coal dry it is subject to spontaneous combustion.

Gasoline – Chevron states that its gasoline can be stored for a year without deterioration when storage conditions are good. That means a tightly closed container and moderate temperatures.

It is recommended that you use a federal, state, and locally approved metal container and avoid plastic.

Only fill the container 90% full. This allows for expansion in warm weather.

Cap containers tightly.

Store containers out of direct sunlight in a location where the temperature stays below 80°F most of the time. Storage temperature affects storage life.
Always use a fuel stabilizer like Sta-Bil or Pri-G. Fuel stabilizer additives are available at most auto supply stores. Follow the instructions on the can.
Personally, I prefer Pri-G since it works and can restore degraded fuel.

Gasoline should be stored in a safe location, never in the house or attached structures.

Diesel - Under normal storage conditions, diesel fuel should have a shelf life of 12 months or longer at an ambient temperature of 68 degrees Fahrenheit.
At an ambient temperature of 86 degrees Fahrenheit, shelf life will shorten to 6-12 months.

Diesel is harder to ignite than gasoline, but remember accidents do happen. Be careful. Two grades are currently available: #1 diesel that is similar to kerosene, and #2 diesel that is similar to #2 home heating oil.

Diesel fuel presents its own unique storage problems. The first is that it is hygroscopic; that means it will absorb moisture from the air. The second problem is sludge formation. Sludge is the result of anaerobic bacteria living in the trapped water and eating the sulfur in the fuel. Untreated, the sludge will grow until it fills the entire tank, ruining the fuel. Stored diesel fuel should be treated with a Pri-D or diesel Sta-Bil. Stored diesel should be filtered before use.

Propane - LP gas is one of the easiest fuels to store and also one of the most dangerous. It is a highly versatile fuel that can be used to power internal combustion stationary engines, tractors, and other motor vehicles, as well as for cooking and heating. LP has serious drawbacks. It must be stored under pressure to remain a liquid. Any leak (which may not be visible) could leak away all of your fuel without your knowledge. In case of a leak the gas, which is heavier than air, will flow like water and puddle in low spots, waiting for an ignition source. Always put your propane tank downhill from your house.

The big advantage of propane is that in most of the country it is readily available. In my area, it is common to see 250 and 500 gallon or bigger tanks in people's yards. It is delivered to your home on a regular schedule.

Small five to thirty gallon tanks are cheap, available and can be connected to Coleman stoves and lamps as well as many barbeques, travel trailers and campers.

Safety Tips

Be sure flammable liquid products are stored away from heaters, furnaces, water heaters, ranges, and other gas appliances. Beware of pilot lights.

Flammable liquids should be stored out of the reach of children.

Flammable liquids produce invisible, explosive vapors that can ignite by a small spark at considerable distances from the flammable substance. Store them outside the house.

Keep at least a 4A 10BC rated fire extinguisher available near the flammable liquid storage.

Always obey federal, state and local fire regulations, and use approved safety storage containers for the storage of flammable liquids.

Chapter 11- Medical Supplies

This chapter is not a first aid manual. It is simply a list of supplies that I suggest you store. There is a list of over-the-counter medicines. These should only be used as per the directions on the labeling. Any variation from these directions should only be done at the directions of competent medical personnel.

I have mentioned antibiotics, because I am aware that some are storing veterinary or fish antibiotics. These have not been approved for use on humans. Do not use these without competent medical advice.

The following list was prepared with the advice of a doctor. It is a list of first aid supplies. Depending on your medical training, or upon the advice of your doctor, you may want to add additional medical supplies.

The quantities of first aid supplies required for your family will depend on its size. This is not a complete list; contact your own medical provider for advice on any of your special medical needs.

Abdominal pad 5" x 9", minimum of 6
Absorbent gauze pads 2x2, 4x4
Adhesive tape (1/2", 1", and 2")
Airway, pharyngeal, plastic, adult, and child sizes
Antiseptic, such as Betadine (povidone) or Hibiclens
Assorted Band-aids
Bandage roller, assorted sizes
Blade, surgical knife, sizes 10, 11, and 15, one-half

Dozen each
Blood stopper kit
Burnfree Dressings
Butterfly bandages
Cotton, absorbent, sterile
Cotton tipped swabs
CPR mask
Ear syringe
Elastic bandage 3", minimum 2
Eye pads
Forceps, 6" assorted, both straight and curved
Handle, scalpel #3, for detachable surgical knife blades
Gloves, disposable vinyl, nitrile or latex gloves, minimum of one box
Hypodermic syringe (60 CC)
Multi Trauma dressing, 12" x 30"
Petroleum jelly
Advanced Red Cross first aid book
Roller gauze 1" and 2"
Respirators N-95, minimum level of respiratory protection is a surgical mask or preferably a N95 respirator
Safety pins
Scissors, bandage, and straight
Sphygmomanometer, aneroid (blood pressure cuff)
Splints two 36" SAM Splints plus two 4" rolls of
 Cohesive wrap
Sponges 4" X 4" Sterile, minimum 10
Stethoscope
Surgical soap
Sutures, assorted
Suture holder

Thermometer, old style mercury thermometers, both oral and rectal
Tongue depressors
Tourniquet
Triangular bandage 40" x 40" x 56"
Tweezers
Vaseline

Non-prescription medication

Alcohol, rubbing
Antacid
Anti-diarrhea medication
Aspirin, Tylenol or Excedrin
Baking soda
Benadryl
Calamine lotion
Eyewash
Gatorade powder, used for dehydration. If you cannot get Gatorade you can make your own rehydration solutions. Mix 8 level teaspoons of sugar, ½ teaspoon of salt and one liter of purified water.
Hydrocortisone ointment
Hydrogen peroxide
First Aid cream
Laxative such as Senekot
Multi-vitamin and vitamin C
Salt
Sunscreen

Store your required prescription medications as needed. Freezing will extend the shelf life of some medications. Consult your own doctor or pharmacist.

Remember that preventing an infection is better than treating one. Clean any wound with clean water and soap. Do your best to wash any dirt or debris out of the wound. Irrigate the wound with clean water. You can use a large hypodermic syringe (**60** CC) without the needle to squirt water under pressure to clean the wound. If you lack a hypodermic syringe, you can use a plastic bag. Fill it with water, cut a pinhole in a corner, and squeeze. Use a good antiseptic or antibiotic ointment if you have them.

A good medical pack that I recently ran across is the M3 Medics Bag. The one I obtained from Freezedryguy was filled with a well thought out list of medical supplies.

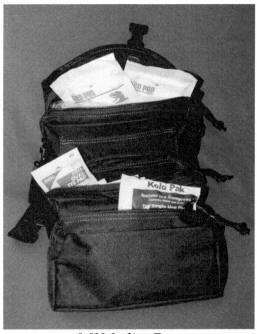

M3Medics Bag

Are over-the-counter (OTC) drugs really worth stockpiling? Remember most of the OTC drugs on this list were prescription medications when first released. Although many other OTC drugs are worth considering, these have been selected due to their ready availability, affordability, safety in both adults and children, and multi-use potential. Used alone or in combination, they can effectively treat dozens of conditions including headache, fever, sore throats, earache, heartburn, arthritis, ulcers, diarrhea, allergies, hives, congestion, dizziness, mild anxiety, nausea, vomiting, poison ivy, athlete's foot, ringworm, eczema, insomnia, backache, gout, diaper rash, yeast infections, and many common illnesses.

Over the counter medications are serious medicines and should not be misused. Read and follow the directions on the packaging.

Acetylsalicylic acid (Aspirin) - Aspirin is used to reduce fever and to relieve mild to moderate pain from headaches, menstrual periods, arthritis, colds, toothaches, and muscle aches. Aspirin is also used to prevent heart attacks in people who have had a heart attack in the past or who have angina (chest pain that occurs when the heart does not get enough oxygen). Aspirin is also used to reduce the risk of death in people who are experiencing or who have recently experienced a heart attack. Aspirin is also used to prevent ischemic strokes (strokes that occur when a blood clot blocks the flow of blood to the brain) or mini-strokes (strokes that occur when the flow of blood to the brain is blocked for a short time) in people who have had this type of stroke or mini-stroke in the past.

In general, aspirin works well for dull, throbbing pain; it is ineffective for pain caused by most muscle cramps, bloating, gastric distension, and acute skin irritation. Throw away any aspirin that smells like vinegar.

Ibuprofen (Motrin, Advil) - Among the OTC anti-inflammatory medications, ibuprofen is probably the most versatile. Nonprescription ibuprofen is used to reduce fever and relieve pain of headaches, muscle aches, arthritis, menstrual periods, the common cold, toothaches, and backaches. It is also effective at reducing fever and is generally safe for use in children. It is not advisable for most stomach-related pain, although it may decrease the pain of kidney stones, kidney infections, and possibly bladder infections. The most common side effect is stomach irritation or heartburn. When combined with acetaminophen, it is nearly as effective as codeine, tramadol, or hydrocodone in relieving more severe pain.

Warning: I would not recommend combining any medicines without the advice of a licensed physician. Combining any medication can cause severe injury or death.

Acetaminophen (Tylenol) - Acetaminophen is the only OTC pain-reliever that is not an anti-inflammatory drug. It will not irritate the stomach like ibuprofen, aspirin, or naproxen. Acetaminophen is used to relieve mild to moderate pain from headaches, muscle aches, menstrual periods, colds and sore throats, toothaches, backaches, and reactions to vaccinations (shots), and to reduce fever. As mentioned above, it may be combined with ibuprofen in full doses for more severe pain. Side effects are very few, though in high doses, especially when combined with alcohol, it can lead to liver failure. It is available in several pediatric dosages, both for pain relief and fever reduction.

Diphenhydramine (Benadryl) - An inexpensive antihistamine, diphenhydramine is primarily used to relieve red, irritated, itchy, watery eyes; sneezing; and runny nose caused by hay fever, allergies, or the common cold. Diphenhydramine is also used to relieve coughs caused by minor throat or airway irritation. It may be used to prevent and treat motion sickness. It is also indicated for hives and itching, including itchy rashes such as poison ivy. Although not all patients become drowsy when using diphenhydramine, many do so, making this medication useful for insomnia as well.

Loperamide (Imodium) – Loperamide is the most effective OTC medication for diarrhea. It is available in both a tablet form and liquid for children.

Contraindications: Loperamide should not be used if the body temperature is over 101degrees F (38C) and in presence of bloody diarrhea.

Pseudoephedrine (Sudafed) -Pseudoephedrine is used to relieve nasal congestion caused by colds, allergies, and hay fever. It is also used for the temporary relieve of sinus congestion and pressure. Pseudoephedrine will relieve symptoms but will not treat the cause of the symptoms or speed recovery. Pseudoephedrine is in a class of medications called nasal decongestants. It frequently has a stimulatory effect, similar to caffeine. The most common side effects are those resembling a burst of adrenaline: rapid heart rate, palpitations, and increased blood pressure. Years ago, this drug was used in young children, even babies, though now most pediatricians do not advise it in patients younger than about six years old.

Meclizine (Bonine, Dramamine) - Meclizine is used to treat or prevent nausea, vomiting, and dizziness caused by motion sickness. It is also used for vertigo (or dizziness) caused by certain inner ear problems. For some patients it causes drowsiness, and therefore may be used as a sleep aid.

Ranitidine (Zantac) - Although several medications are available OTC for the treatment of heartburn, ulcers, and other acid-reducing conditions, ranitidine is among the best-tolerated, is inexpensive, and is also useful for relieving hives. Doctors often recommend an acid-reducing medication such as ranitidine for patients who experience stomach upset when taking ibuprofen, though this must be done with caution.

Hydrocortisone cream - The 1% version of hydrocortisone is the strongest steroid cream available over the counter. It is safe for use in both adults and children in treating inflamed and/or itchy rashes such as eczema, poison ivy, diaper rash, and other minor genital irritations.

Bacitracin ointment - This ointment is best used to prevent skin infections when the integrity of the skin has been breached, as by an abrasion, laceration, insect bite, or sting. It also may be used to treat a superficial skin infection such as a mildly infected wound or impetigo. It cannot be used to treat deeper infections, which generally require an antibiotic by mouth.

Clotrimazole (Gyne-Lotrimin) The same antifungal medication, clotrimazole, is contained in both Lotrimin and Gyne-Lotrimin. Gyne-Lotrimin may be used to treat both female yeast infections and any other yeast or fungal infection that Lotrimin would treat, including athlete's foot, jock itch, ringworm, diaper rashes, and skin fold irritations.

Guaifenesin (Musinex) is used to relieve chest congestion. Guaifenesin may help control symptoms but does not treat the cause of symptoms or speed recovery. Guaifenesin is in a class of medications called expectorants. It works by thinning the mucus in the air passages to make it easier to cough up the mucus and clear the airways.

Gatorade powder: Electrolytes, particularly sodium, are critical for proper hydration, helping maintain electrolyte balance, and helping your body hold on to the fluid it needs. Gatorade can be used to hydrate sick or injured persons who have lost body fluids.

An improvised rehydration solution can be made by mixing half of a level teaspoon of salt, and 8 level teaspoons of sugar in a liter of water. A second variation is to put a half teaspoon of salt, 8 heaping teaspoons of a finely powdered cereal (powdered rice is best) and 1 liter of water. Boil for 5 - 7 minutes to form a watery porridge. Cool and give to patient to drink.

A number of people have suggested the storage of antibiotics. Talk to your doctor and see if he will help. However, realize that without real medical professionals most of us would just be guessing about diagnoses, type of antibiotics required, and doses.

Warning: I do not advocate the storage of any prescription medication without a prescription. I strongly recommend that you only use antibiotics upon the directions of a licensed physician. Misuse of antibiotics can cause severe injury or death.

I am aware that some people are storing fish antibiotics and claim that they are identical to those used to treat humans. They are available through some internet veterinary suppliers. They are marked for fish use only and I cannot recommend them.

It must be stressed that the unsupervised lay use of antibiotics is dangerous for several reasons.

1) Antibiotics may cause potentially fatal reactions (e.g., allergy, asthma, and death).

2) Antibiotics can prompt greater growth, development, and spread of resistant pathogens such as fungi and Mycoplasma prompting more severe or alternative infections.

3) Antibiotic usage can make it more difficult for physicians to diagnose life-threatening infectious illnesses. Thus, self-medication is not advised.

The above information and list of medical supplies is only as good as the person utilizing them. Take an advanced First Aid class, learn CPR, or consider becoming an EMT. There are some good medical books listed in the Reference Section.

Chapter 12 - Sanitation

This is a subject not to be taken lightly. During all the wars, the United States fought prior to World War 2, more men died of disease than combat. At times, diseases caused two deaths for every one killed in combat. This was a result of bad water and unsanitary conditions. Water has been addressed in Chapter 3; other sanitary issues will now be discussed.

Disease and human waste - Today we are lucky to live in a world of flush toilets and plenty of toilet paper. Tomorrow we may have to live like our ancestors. The disposal of human waste has been a major cause of disease for most of human history. The exception has been the last 100 years.

I have known friends who have large food storages, but have not considered hygiene. Your need for sanitation supplies will vary depending on the number of individuals using your facilities and your location.

Items That I recommend Storing are:

Personal – Tooth brushes, dental floss, combs, nail clippers, shaving gear (men), hand clippers, or scissors for cutting hair.

Hand soaps - Bar soap keeps better than the new liquids, which can evaporate, if stored in the heat. Figure out how much hand soap you use in a given period and double the amount. You will probably be dirtier than normal.

Hand sanitizer – With the availability of all the new hand sanitizers, it is a good idea to keep some on hand. They are very convenient when water is in short supply. As an additional use, they are high in alcohol content and make a good fire starter.

Showers - A friend of mine takes a one gallon garden sprayer with him when he goes camping. He uses it to take showers with. It works very well. If it is a sunny day, set the sprayer full of water in the sun and you will have a warm shower.

Remember, a black water container set in the sun will provide you with warm water on sunny days. Whenever I go camping, I set a black rubber five gallon container of water on the roof of my car and it heats up quick.

Dish soap - liquid, or powder. The soap should be biodegradable because you may have to use the grey water for your garden. Look for soaps that will work well in cold water.

What do you do when the soap and scouring pads run out? Scouring pads can be replaced with clean sand. Rub the inside of your pans with sand and a little water. This will remove caked on grease and food particles. It will not remove fats and oil from your dishes.

Fats and oils can be removed by using wood ash. Obtain pure wood ash. Do not use ash from a fire in which you or others have burned plastics or garbage. Scrub the dishes with the wood ash and water. You can heat the water by adding hot coals. Be sure the water is boiled, treated, or filtered. Do not wash or rinse your dishes in contaminated water.

The lye from the wood ash can dry out the skin on your hands if not promptly rinsed off.

Laundry soap - It should be biodegradable; you may have to use the grey water for your garden. Consider putting in a clothesline. Do you have a washboard and tub? Some of the suppliers listed in the Reference Section have hand wringers and other alternatives to washboards, etc.

My daughter has recently started making her own laundry soap. She says it is much cheaper than purchasing it at the store and she likes it better. Her recipe is simple: one bar of Fels-Naptha, one cup of borax and one cup of washing soda. Finely grate the Fels-Naptha and mix with the borax and washing soda. She uses 1/8 cup per load. This may vary depending on your machine.

Washing soda is different from baking soda. You can find it with the powdered laundry detergents in most grocery stores. Arm & Hammer brand is the most common.

Making soap

Lye - First, you have to make lye. You need a wood or plastic barrel (do not use metal). The barrel can be open at one end. It needs a drain at the other end. This can be a hole plugged with a cork or wooden stopper. Stand the barrel on end with the open end up. Make sure the hole in the bottom is plugged tightly. Place some fist size rocks on the bottom of the barrel. Cover the rocks with straw, grass, or hay. Fill the barrel with pure wood ash (burned wood only, no garbage).

Add enough clean water to cover the wood ash. As the ash settles, you may add additional ash. Let it sit for at least three days. Drain the liquid from the barrel and check to see if the lye is finished. Drop a potato or egg into the lye.

If the potato or egg floats with an area the size of a quarter showing, the lye is ready. If the lye is too weak, pour the liquid back into the barrel. Let it sit for a day and reprocess it.

Warning: Lye, whether you make it yourself from wood ash or purchase it at a store, is very irritating to the skin and can do severe damage to eyes and throat. Always use extreme caution when handling lye. Keep it away from children. Rubber gloves and safety glasses should be worn when handling lye.

Warning: Never pour water into lye. Always add lye to the water. If you pour water onto lye, it can cause a violent reaction.

Lard was the most commonly used fat to make homemade soap. Lard was normally obtained from rendering animal fat and waste cooking oil. The fats and oils are placed into a large kettle and an equal amount of water added. This was boiled over an outside fire because the smell was too strong to do it indoors. The mixture is boiled until all the fats are melted. The mixture is removed from the fire and more water is added. This amount of water should be approximately equal to the first amount. Allow the mixture to cool over night. The pure fat is solidified and floating on the top. Remove this fat from the mixture.

Mixing soap - Mix 1 part fat, two parts water, and three parts lye in a kettle (remember to add lye to water). Bring the mixture to a boil and then reduce to a simmer. The soap will float to the top. Skim the soap off and let it cool. Place the soap in a clean pot with two-parts water and bring it to a boil, reduce to simmer. The finished soap will float to the surface. Skim the soap off, let it cool, and cut into blocks.

Although lard is the main ingredient in soap, one can successfully substitute other oils. Substitutions for lard can be sunflower, canola, olive, or vegetable oil

The utensils you use in soap making should be saved for soap making use only and should never be used for cooking or preparing food. This includes the kettle you use to cook the soap.

You must **never** use pans and utensils made from aluminum, iron, tin, or Teflon for soap making. You can use cast iron, stainless steel, enamel coated steel or a heat resistant glass container like Pyrex.

Warning: Keep lye out of the reach of children and animals. Lye is very caustic and can cause serious injury or even death if swallowed and can cause blindness if splashed into the eyes. Wear long sleeve shirts, gloves, and safety goggles.

Disinfectants such as Pinesol, Lysol and Hexol, are good to have on hand in case of illness. Bleach is a good disinfectant; it is inexpensive and purifies water. Nevertheless, it needs to be rotated. It loses half its strength in one year.

Feminine supplies - How much you stock depends on your family's needs, but be generous. The individually wrapped feminine napkins make good first aid dressings. In the past, women have used rags and pads of moss.

Diapers - Be old fashioned and stock cloth diapers. After the baby out grows them, they make good rags. Buy what you think your family will need.

Remember, there may be no modern methods of birth control available.

Toilet paper - Stock what you consider practical. However, remember that for most of human history, toilet paper was not available. While it seems like a necessity in today's world, many other items have a higher priority for your survival. In some South American countries, they use sanitary washcloths. You will need three to five washcloths a day for each family member. After they are used, place them in a closed container full of soapy water. The cloths need to be washed once a day and dried in the sun. The sun helps to sanitize them.

Old magazines, corncobs, and leaves have all been used in the past in place of toilet paper. Be sure you avoid poison oak, poison ivy and other irritants.

Dental hygiene - The condition of your teeth affects your overall health. Keep flossing and brushing. An improvised toothbrush can be easily made. Take a twig from a tree (willow or poplar is best). Sharpen one end to use for picking your teeth. Chew on the other end and use the fibers as a brush.

Sun washing clothes - If you have a real water shortage, shake your clothes out and spread them out in the full sun. The more the clothes are exposed to sun, the better. Sun washed clothing will feel cleaner and smell better. The ultraviolet radiation will kill off the bacteria that live in your sweat and dead skin cells. Do not forget to sun wash your sleeping bags and bedding.

Human waste - A temporary toilet can be made out of a five-gallon bucket and one of the toilet seats from your home. If you have plastic bags to line the bucket with, it is easier to dispose of the waste. Keep a can of wood ash, lime or other disinfectant by your improvised toilet to sprinkle over your waste. This helps to keep the smell down and the insects away. Cover the toilet when it is not in use. The waste can be disposed of by burying or burning. The problem with burning is that it takes a lot of fuel. Burying works well, but you have several things to consider.

How close is the nearest water? The U.S. Forest Service advises you to be at least 200 feet from a stream or well when you bury your waste. Consider the slope of the land and, if possible, bury the waste down slope from your water source. Remember, when it rains, you do not want the water to drain through your waste disposal area into your water source. When you bury waste, cover it with at least 18 to 24 inches of dirt.

If you are without normal sewage facilities for a long period, you may have to improvise them. Dig a slit trench 18 inches wide, 4 feet long, and at least 4 feet deep. Rig up a privacy partition around it. You can use almost anything from brush, scrap lumber or old sheets, etc. Since you will have to squat over the trench, you will need something to hang on to. Install a couple of posts within easy reach, you can even run a rope between them for additional support. Provide something to wipe with and wood ashes to cover the waste and you are in business.

Recently, I had a problem with my plumbing late in the afternoon, the sewer plugged solid. My toilets were out of service. Now this is something I had prepared for. I have all kinds of plans and supplies for long-term use. However, at eight o' clock in the evening on a cold night, I did not feel like digging though a storage area to set up a toilet that is designed for long term use. I only needed it until morning.

I realized that for short-term emergencies or for a transition period into a long-term situation, an inexpensive camping toilet or an old fashioned pot is very convenient.

Worms - These are something that most of us have not had to face in recent years. In many foreign counties, they are still quite common. Poor sanitation is the biggest reason for their spread. Worms are spread when people relieve themselves on the ground. The feces may contain roundworm eggs. Even if someone cleans the area, some eggs may remain in the ground. Roundworm eggs can remain infectious for many months. When children or adults contact the contaminated soil, the worms may infect them. Young children are particularly susceptible, since they often play or crawl on the ground.

Rodents and insects - Fleas, ticks, cockroaches, bedbugs, tapeworms, rats, and mice are problems that our ancestors dealt with on a daily basis. They all are potential disease carriers. In a long-term emergency, they would become our problem. In planning your storage, you need to consider these issues. Your storage should include old-fashioned rat and mousetraps, rat poison, insect sprays and repellents. In addition, Chapter 15 Miscellaneous Recipes includes some old-fashioned insect repellants.

Flies are also responsible for the spread of disease. They like feeding on feces and can travel long distances. Flies have spikes on their legs, so particles of whatever they feed on are carried with them. If the flies are feeding on the feces of someone suffering from a diarrhea disease such as gastroenteritis, these particles may pass the disease on to others. The fly's next meal is quite likely to be on human food. So bits of feces are left behind on food or drink which is then eaten by people. The disease is then passed on.

Many of us routinely use air conditioning and do not often open our windows, so our window screens are not well maintained. If you intend to shelter in place consider maintaining your screens as part of your emergency preparations and do not forget the fly swatters.

In a survival situation, hygiene is more important than normal; you cannot afford to get sick. Wash whenever you can and keep your food preparation areas as clean as possible. Keep your dishes, pots, and pans clean. Clean sand can be used as an improvised scouring powder.

Washing Clothes by hand - is very hard on your hands. A very inexpensive way to wash clothes that is not hard on your hands is easy to improvise. All it takes is a five-gallon bucket and a new toilet plunger. Use the toilet plunger to agitate your clothes.

Mobile Washer manufactures a type of plunger specifically for this purpose. While it costs a little more, it does create a better up and down action in the water.

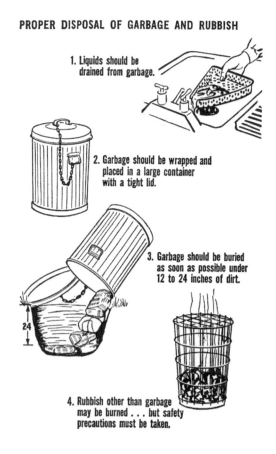

PROPER DISPOSAL OF GARBAGE AND RUBBISH

1. Liquids should be drained from garbage.
2. Garbage should be wrapped and placed in a large container with a tight lid.
3. Garbage should be buried as soon as possible under 12 to 24 inches of dirt.
4. Rubbish other than garbage may be burned . . . but safety precautions must be taken.

Chapter 13 - 72-hour kits

Everybody from The Office of Homeland Security to the American Red Cross advises you to have a 72-hour kit. They all have lists of suggested supplies. After reading their lists, I feel that they are all on the weak side. I operate on the assumption that whatever happens will occur at midnight on the coldest, wettest night of the year. If you are prepared for this, anything less will be easy.

Before you start on your kit, I want you to stop and think about how much you intend to spend. A common excuse is, "I cannot afford it." If that is your excuse, look at the next homeless person you see. How much do you think his gear cost him? It may not look fancy, but it is keeping him alive. He is not just looking at 72 hours; his kit is his whole life. Do not hesitate to improvise. Go to flea markets, thrift shops, or garage sales. Some great equipment has been found in them inexpensively.

Each family member should have his or her own kit. They should be personalized for their needs. Kits should be kept at a central location that everyone is familiar with. Inspect your kit at least twice a year. Check children's clothing for proper fit.

A 72-hour kit needs to meet your four most important needs: food, water, shelter and fire. Other than special medical requirements, these are the most important requirements to keep you alive.

To minimize weight try and find items that serve multiple purposes and avoid duplication.

Three day's supply of food - Your needs will be determined by the size and age of your family and their personal needs. Remember that it always helps to have extra food. If you have to utilize your kit, you will be under a lot of stress. This is not a good time to skimp on food. You should plan for 2000 to 3000 calories a day. Many of the commercial kits currently being sold only provide 800 to 1200 calories a day.

MREs - Suggested items include MREs (Meals Ready to Eat). MREs are a self-contained, individual field ration in lightweight packaging bought by the United States military for its service members for use in combat or other field conditions where organized food facilities are not available. MREs have a good shelf life if kept in a cool spot. They are a complete meal including main course, dessert, snacks, drink powder, and plastic utensils. Except for the beverages, the entire meal is ready to eat. While the entree may be eaten cold when necessary, it can also be heated in a variety of ways, including submersion in hot water while still sealed in its individual entree package. Each meal provides about 1,200 calories.

I do not recommend that you eat MREs for more than a few days. They are low in fiber and have a tendency to cause constipation.

I have recently received information on a new shelf life chart being used by the Defense logistics agency. This chart shows a shorter shelf life than the previous chart. The old chart is widely published on the internet. This chart was developed in the 1980's when the MREs first came out and projects up to 130 months of storage at 60 ° F. These first MREs contained a number of freeze dried components. Today's MREs no longer contain the freeze dried components and there have been other changes to improve the taste. This accounts for the shorter shelf life.

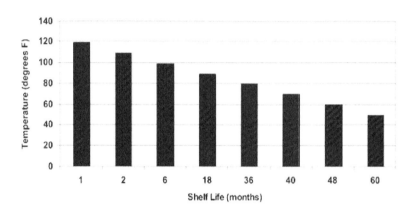

Defense logistics agency - How long will it last?

"The shelf life of the MRE is three (3) years at 80 degrees F. However, the shelf life can be extended through the use of cold storage facilities prior to distribution." The previous statement is taken directly from the current website of the defense Logistics agency and confirms the information given on the new chart.

But do not despair if you have a stack of the older MREs; I have eaten the old ones that were over 15 years old. The taste was not quite as good, but they did not kill me. They had been stored under good conditions.

My own feeling is that the majority of the MRE components will probably last considerably longer. But some MRE components might not fare so well over time.

More about MRE Shelf Life

1. The shelf life ratings shown in the chart on the front of this paper were determined by taste panels (panels of "average" people, mostly office personnel) at the Natick lab. Their opinions were combined to determine when a particular component, or in this case the entire MRE ration, was no longer acceptable.

2. The shelf life determinations were made solely on the basis of taste, as it was discovered that acceptable nutritional content and basic product safety would extend way beyond the point where taste degradation would occur. This means the MRE's would be safe and give a high degree of food value long after the timing suggested in the chart.

3. MRE pouches have been tested and redesigned where necessary according to standards much stricter than for commercial food. They must be able to stand up to abuse tests such as obstacle course traversal in field clothing pockets; storage outdoors anywhere in the world; shipping under extremely rough circumstances; 100% survival of parachute drops; 75% survival from free failure drops; severe repetitive vibration; and individual pouches being subject to a

static load of 200 pounds for three minutes.

4. Freezing an MRE retort pouch does not destroy the food inside, but repeated freezing increases the chance that the stretching and stressing of the pouch will cause a break on a layer of the laminated pouch. These pouches are made to withstand 1,000 flexes, but repetitive freezing does increase the failure rate by a small fraction of a percent. Also, if MRE food is frozen, then thawed out, it must be used the same as if you had thawed commercial food from your own freezer at home.

Time and temperature indicator
Since about 1997, military MRE cases have also included something called a TTI (time and temperature indicator) on the outside of the box to assist inspectors in determining if MREs are still good. There are two parts to the TTI - an outer dark circle and an inner light circle. As long as the inner circle is still lighter than the outside circle, the MREs are supposed to be good. The following picture shows an example of a TTI:

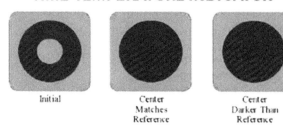

Each MRE weighs 18 to 26 ounces, depending on the menu. Since MREs contain water, they weigh more than freeze-dried meals providing equivalent calories. If you decide to carry MREs, you may want to break them down into their components and throw the outside packaging away, to lessen the weight. Even "stripped down" to their essential packaging (throw away their plastic pouch, cardboard containers and anything else you do not want), they are still bulkier and heavier than dehydrated and freeze dried foods.

Some people hang on to a few of the outer plastic bags to use for gathering water. The cardboard is also good as a fire starter.

There are currently two types of MREs. The government has made it illegal to sell military MREs. The ones being sold through the commercial outlets are of varying types. All use some MRE components. These are purchased from the same companies that manufacture them for the military. Civilian companies put them together in different fashions - some to military specs, some to their own. The military ones have MRE heaters, the civilian ones often do not. If you buy, MREs make sure they meet military specs. Many of the civilian MREs do not meet the standards required by the military.

Commercial freeze-dried or dehydrated, individually packaged meals are available at most sporting goods stores and many discount houses. The plus side is that they are convenient, nutritious, tasty, and lightweight.

The downsides are they require water to prepare and they are expensive.

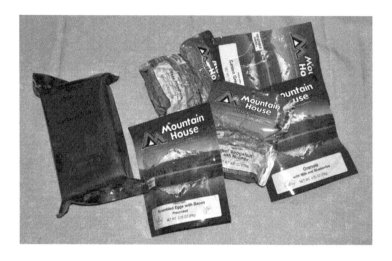

Military MRE on left, Mountain house meals from their "Just in Case" Unit on right.

The "Just in Case" Unit by Mountain House consists of 21 meals, including breakfasts, lunches, and dinners. It provides all the food for a 72-hour kit for two people.

The Mountain House meals have a seven-year shelf life at 75 degrees or less and taste great.

Canned food – meats, tuna, sardines, fruits, vegetables, chili, etc., all are good choices. The downside is their weight and the need to rotate.

Top Ramen Noodles - They need to be rotated often and are not well packaged (bugs can get in). Pack them in plastic Ziploc bags. They require water to prepare. The plus side is they are light and cheap.

Granola Bars or Power Bars, New Millennium Energy bars, hard candy, fruit bars or fruit rolls, dried fruit and trail mix, etc., are all good products. However, you have to remember to rotate them, particularly anything with nuts. The oil in the nuts will turn rancid.

Lifeboat rations, or ration bars, provide approximately 1200 calories per day. Many of the prepackaged 72 hour kits contain these types of bars. They are sold in 2400 and 3600 calorie packages and can be broken into 400-calorie sections. A meal is one 400-calorie section. Their taste is not too bad. Depending on the brand, they taste like lemon butter cookies or shortbread. The bars consist mainly of flour and sugar fortified with vitamins.

The rations will keep you alive, but you will be hungry and probably unhappy. Still, there are some advantages to lifeboat ration bars. They will withstand long periods of heat and freezing temperatures. The bars are compact, lightweight, and packaged to withstand rough conditions. Because they are small and so inexpensive, I always double the rations.

If you are making a kit that will be left in your vehicle, lifeboat rations are the best solution. They have a five-year shelf life and will withstand long periods of high temperatures or sub-freezing weather without significant deterioration.

Cooking utensils - My personal choice is the stainless steel U.S. military surplus mess kit. They are inexpensive, sturdy, and reasonably lightweight. The US stainless steel knife, fork, and spoon set make a good companion. There are many good commercial backpack cooking sets available at most sporting goods stores for a bit more money. It is recommended that you avoid aluminum cooking sets and plastic silverware. Plastic silverware is useless for cooking over an open fire.

Water -A good combination is a US military surplus 1-quart canteen and cover. As an accessory, you can purchase a military surplus cup and a small stove. Both will fit inside the canteen cover. The canteen nests inside them. The stove works with military heat tabs that are readily available. One heat tab will boil a cup of water. Canteen covers have a pocket for water purification tablets. Carry at least 2 canteens. These components can be purchased at most military surplus stores and many of the suppliers listed in the Reference Section.

There are many good commercial water bladders, canteens and hydration backpacks on the market. Just be able to carry at least 2 quarts of water and have a method to purify more. If you live in a desert area and are physically able, you may want to carry 4 quarts or more of water. In an emergency canteens can be improvised from water or juice bottles stuffed into whatever bags you have available.

Many of the water bottles currently on the market are made of stainless steel. These have an advantage over plastic in that you can boil water in them.

From left to right: 1-quart US canteen, bottle of water purification tablets, stove and canteen cup with heat tabs in front, canteen cover. Canteen fully assembled includes everything but the heat tabs (fuel). The purification tablets fit in the pouch on the right side of the canteen cover.

Water can be purchased in small four-ounce foil packs that are manufactured for use in lifeboats. These have a shelf life of approximately four to five years. If you choose to carry these, remember that you need a gallon of water a day. There is 128 fluid ounces in a US gallon.

Water purification - There are many suitable methods of water purification for a 72-hour kit. Chemical tablets or water filters are best. Boiling or SODIS both work well, but takes too much time if you need to evacuate.

Aquamira, Micropur and Portable Aqua all manufacture water purifications tablets containing chlorine dioxide that kills bacteria, viruses, and cysts, including Giardia and Cryptosporidium. Chlorine dioxide does not discolor water, nor does it give water an unpleasant taste like iodine. They have a four to five year shelf life; check the expiration date on the packaging. While chlorine dioxide is the best product on the market, it has one downside. Some of the tablets require a 4-hour reaction time prior to drinking. Chlorine dioxide water treatment drops manufactured by Aquamira only require a fifteen-minute reaction time.

Portable Aqua iodine tablets come 50 per bottle. They have a 4 year shelf life but only 6 months after the seal is broken. One tablet per quart is a minimum water treatment. It is inexpensive and can be found in the camping section of almost any large discount or grocery store. Iodine will leave a taste in the water.

Polar Pure is an iodine crystal based product. It is more expensive and harder to find than Portable Aqua (check the suppliers in the Reference Section). The advantage of Polar Pure is that it has no shelf life, and according to the instructions, a bottle will disinfect 2000 quarts of water. The disadvantage is that if you consume that much iodine you may have medical complications.

Povidone 10% solution, normally sold under the name Betadine, is commonly found in most first aid kits. It will disinfect water; use 10 drops per gallon in clear water, up to 20 drops in turbid water. Portable Aqua Iodine tables, Polar Pure, and povidone cause a slight taste in the water. The taste will not hurt you. Povidone does contain iodine.

Warning - Pregnant or nursing women or persons with thyroid problems should not drink water disinfected with iodine. Prolonged use of iodine can cause medical problems in some people. In addition, some people who are allergic to shellfish are allergic to iodine.

Warning - Iodine will not reliably kill Giardia and Cryptosporidium.

Water filters - First Need, Katadyn, and Aquamira make excellent water purifiers. They have a large selection of sizes and price ranges. Choose the size that fits your family. The Frontier Pro by Aquamira is currently being used by the US Military. It is small, compact, efficient and reasonably priced. There are other brands on the market that are just as good, but I have tried these three brands and can recommend them. If you decide to purchase a different brand, make sure it will eliminate Giardia and Cryptosporidium. Many of the products on the market are filters and not purifiers; this means that they may not filter all dangerous bacteria out of the water. Read the specifications.

SteriPEN is another method of water purification that you should consider. It uses UV light to purify drinking water. Both battery and hand crank models are available.

Clockwise from top left: Povidone 10% solution (Betadine), Portable Aqua Tablets, Water purifier Tablets and Water treatment Drops by Aquamira, and Polar Pure

Clockwise from top left: Katadyn Water Purifier, Frontier Pro by Aquamira, and a First Need water purifier.

Fire starting - Matches and cigarette lighters are the fire starters most people are familiar with. They are readily available and every home probably has a few books or boxes of matches stuck in a drawer somewhere. Matches are handy, cheap, store well and are easy to use.

The problem is not with the matches, it is with us. We all like to think we can start a fire with one match. Try it; most people will be surprised at how hard it is to start a fire in good conditions with dry wood. Starting one on a cold windy night with wet wood takes practice. Most waterproof match cases only hold about a dozen strike-anywhere matches. Store extra matches; they will be good to trade with those who do not practice fire starting. I recommend the strike-anywhere kitchen matches.

Store your matches in a waterproof container and rotate them regularly. The more often you carry them, the more they need to be rotated. The movement causes the chemicals on the heads to flake off and the matches may not strike.

Butane lighters will not work in low temperatures after a few minutes exposure. Carry them inside your clothes where your body heat will keep them warm.

Military heat tabs or other prepared fire starters belong in every kit. You can purchase them through the suppliers in the Reference Section or make your own. They can be as simple as cotton balls impregnated with Vaseline. An old 35 mm film can makes a great storage container for them. Another fire starter is to melt wax in a double boiler and add sawdust until the mixture starts to become stiff. Then pour the melted mixture into the bottom of old egg cartons and let it harden. Because of the fire hazard, it is best to do this chore outside. Once the mixture has hardened, you can cut the carton in twelve sections and carry a couple with you. You can break a section up and use as much as needed.

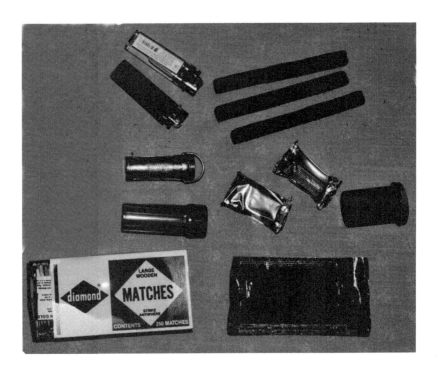

Clockwise from top left: Cigarette lighters, commercial fire starters, film canister with vaseline impregnated cotton balls, military heat tabs, matches, and waterproof match cases.

Clockwise from top left: BlastMatch, Strike Force, and a Speedy Sharp, a Sparkie, magnesium block and flint, and two flints, one on key ring.

There are many commercial fire starters such as the ones consisting of a block of magnesium and a flint. You see this all the time on the television show, "Survivor". They are available in most sporting goods sections. If you take the time to learn to use them they work well, but they do take two hands. Scrape some of the magnesium into your tinder pile before creating sparks. I use the Speedy Sharp for both scraping the magnesium and making the sparks. The Speedy Sharp also serves double duty by sharpening knives.

Innovations such as the Gerber Strike Force and the BlastMatch are improved fire starters. The BlastMatch can be used with one hand. The one that best suits my requirements is the Sparkie from Ultimate Survival. It weighs less than 1 oz and can be used one-handed. I always carry a small flint on my key chain for backup. Do not forget to practice; flints are hard to use.

Backpack Stoves

There are many small backpack stoves on the market. The problem with most of them is fuel; you need to carry a flammable liquid or a propane bottle. The Sierra stove is one of the exceptions; it burns wood. It utilizes a small fan to create a forge like effect. The fan requires one AA battery. You can carry a rechargeable AA battery and a solar charger that only weighs a few ounces. If your flashlight uses AA batteries, they should also be rechargeable.

From left to right: Coleman fuel, propane and a Sierra stove.

Sanitation - If you are forced to evacuate, you may not have much choice about where you end up staying, but you do have some say about the conditions you live in. Keep your space as clean as possible. Keep yourself clean; it will help your morale as well as your health. Be sure and put the following items in your kit:

- Toilet paper - Store it in a plastic bag to keep it clean and dry.
- Bar of soap - Store in plastic soap dish or zip lock bag.
- Towel - Small hand towel
- Feminine supplies as needed
- Medication as needed
- Toothbrush and toothpaste
- Insect repellant - The type may vary depending on where you live.

- Sun Screen - You do not need the additional problem of bad sunburn.
- Shaving gear
- A comb

Warmth and Shelter

Sleeping bags - A good sleeping bag is the best choice for warmth and comfort. There are many good sleeping bags on the market, but the best is probably the Wiggy's. They are a bit on the expensive side, but how much is a good night's sleep worth? Whatever type of bag you choose; here are some tips to help you:

Goose Down is more expensive, lighter, compresses easier, warmer by weight, more durable and long-lived. However, if it gets wet, it is useless. In extreme cold, your body releases moisture as you sleep, so a down bag can get wet from the inside even when it is protected from the outside elements.

Some of the newer insulations such as Lamilite or Polarguard 3D will keep you warm when wet.

Check the stitching; the tubes should overlap so that the stitching does not go all the way through the bag wall creating cold spots.

Make sure the bag has a sturdy zipper and a draft tube along the entire length of zipper.

Consider an outer Gore-Tex or other water repellant shell for your bag. However, be sure that the shell you purchase will breathe enough to allow body moisture to escape.

Mummy bags are largely lightweight, the most efficient for warmth and takes less room in your pack.

Remember, you have to carry it. Think about the weight.

I often see good quality sleeping bags in garage sales for pennies on the dollar.

There are a few things to remember when using a sleeping bag.

One - try not to breath into the bag. This will be very tempting in cold weather. Breathing into the bag will cause moisture to accumulate and you will find yourself with a damp bag.

Two - if possible take off at least your outer layer of clothing; this will help keep moisture out of your sleeping bag.

Three - insulate underneath the bag. You will lose heat to the cold ground. You should have 3 times more insulating value under you than you have on top. This can consist of foam pads or closed-cell self-inflating pads, blankets, piles of newspaper, a piece of carpet, or leaves to help insulate underneath you. Do not use a blow-up air mattress. Air mattresses only increase the amount of air that you need to heat up.

If you cannot get a sleeping bag, the next best choice is a 100% wool army blanket. Wool retains a lot of its warmth when wet. They are a bit on the heavy side, but are inexpensive. Surplus blankets are available for $10-15, and they are always thrift shops.

Space blanket - They are for emergency use, only when you have no other choice. A friend used one as a cover for his sleeping bag in near zero weather. It caused enough condensation to make his bedding very damp. They are small, light and are designed to help your body retain its own natural heat. For them to be effective you must have sufficient body heat. The colder you are when you start to use them the less efficient they will be.

They are completely ineffective in cases of hypothermia. An external heat source is required for you to survive. Remember they will keep you warm, but they will not get you warm.

On the plus side, they make a fair ground cover. They will shed water and make a good windbreak. They can also be used to collect water, create shade or signal for help. If you decide to go with a space blanket, look for the space blanket bivy sack. This is like a sleeping, bag although made of the same material as a space blanket.

Rain poncho - In my opinion, one of your most important survival items is a rain poncho. I like the US military surplus ones; they are sturdier than most civilian models. Check surplus ponchos for leaks. A poncho can keep you and your equipment dry while hiking. With a little light line, they make a good rain shelter over your bed or a good ground cover.

Tents-I have chosen to add a tent to my 72 hour kit. It is not an ultra lightweight tent that will fit into my pack. I keep it with my back pack and would take it with me if I had to leave in my car or another wheeled method (bike, wagon, or golf cart) of travel.

The tent I have chosen is the Eureka Assault Outfitter 4; this is a four-man version of the current Marine Corp Tent. At 13 pounds, it sounds a bit heavy, which is why I would only take it with me if I had wheeled transportation. It is a four season tent and would help you stay warm and dry under extreme conditions.

There are many other good choices in tents from ultra light up. However, do what I did, take your tent out and spend some time in it. Try it in the wind and rain. Learn how to set it up.

Large trash bags - I always shove a couple of trash bags in my 72-hour kit. They can be used for anything from improvising a sleeping bag to a poncho. For a sleeping bag, stuff them with old newspapers or other insulation. Just remember to leave the top open so they can breathe or you will wake up wet from condensation. A rain poncho can be made by cutting a hole for your head and arms and pulling it on like a sweater. If you are wearing incorrect footwear, a piece of plastic between your socks and shoes can help keep your feet warm.

A piece of 8' x 10' plastic makes a good improvised shelter. You can make a good solid anchor by folding a small rock into the corner and tying a light line around it. Put the plastic over a tree branch or a rope stretched between two trees, and then anchor the corners down with the light lines you have tied around the rocks, and you have a shelter.

Miscellaneous

Battery power or solar power radio- Some of the solar and hand crank radios currently on the market are excellent. The newer radios also have a light and will charge a cell phone. A radio provides you with knowledge that might save your life. Check the references for a good source.

Knives -A good, sturdy high quality knife is one of your most important tools. Avoid cheap foreign brands. Either a folding or straight bladed knife will work; the blade does not need to be over 4 inches long. If you can afford it, consider multi tools like the Leatherman Wave. It is my personal choice. Do not forget a small sharpening stone.

Flashlight - You need a good flashlight. There are many good quality LED models currently on the market. Most use AA batteries. Do not forget the spare batteries. Rotate them periodically. An alternative is to use rechargeable batteries and carry a small solar charger. Beware of cheap foreign imitations.

Rope - An excellent choice is a 50 ft. hank of surplus military parachute line (550 cord). This is very strong for its size. It is made up of numerous small strands of heavy thread inside of a fabric tube. The individual threads can be pulled out and used for sewing or repairing equipment.

Shovel - Include a surplus military entrenching tool and cover, or a small shovel for burying waste. The Glock entrenching shovel is an excellent choice and it also includes a saw. There are many cheap imitation entrenching tools on the market that will break with hard use, so go to a reputable supplier.

Backpack -You need a good backpack or other bag to carry your 72-hour kit. There are many good packs, both civilian and military, on the market. Get a pack that fits you, load your gear, and take it out for a good hike (ten or more miles). I personally like a backpack with an external frame. Make sure yours has a good padded waist belt and distributes the weight between your shoulders and hips.

If you cannot afford a good backpack, try looking in garage sales and thrift stores. You can improvise one out of old luggage, duffel bags, daypacks, etc. If you have to evacuate on foot, try not to carry your pack.

Load your pack into the kid's wagon, onto a bicycle, into a wheelbarrow, golf cart or any other wheeled device you can think of. You can always strap it to your back if your carrier breaks or the terrain gets too rough.

Things to consider when buying a backpack (information from Freeze Dry Guy):

- Comfort
- Load bearing capability (how much weight do you have to carry?)
- Cost
- Color
  Ruggedness
- Versatility

Civilian Backpacks

Advantages
- Usually more advanced
- Normally very comfortable
- Lighter than military

Disadvantages
Usually not as rugged as military
- Often much more expensive
- Fewer places to hang gear on outside of pack

- Often times available only in bright colors (do you want to be seen?)

Military Backpacks

Advantages:
- Much less expensive than civilian
- Widely available
- Very rugged
- Subdued colors
- More places to hang equipment on outside of pack
- Generally, more pockets for storing gear, easier to access more items of equipment
- Some packs can be made quite comfortable with certain after market modifications.

Disadvantages
- Often times not as comfortable as civilian packs
- Usually heavier than civilian packs
- Often not as well designed as civilian packs

Clothing - Your clothing should be appropriate to the climate zone in which you reside. They should always include a warm jacket, hat, and gloves. Your clothes should be good quality. Multi layers are the best way to dress. This gives you control of your body heat. Avoid getting overheated or sweating.

In a cold climate, you want at least three layers. The base layer should be polypropylene, silk or wool long underwear. The intermediate layer should consist of fleece, or wool. The outer layer should be windproof/waterproof shell. This is a minimum; the colder it is the more layers.

Cotton is excellent in a desert climate. It is known for its breathability and ability to retain moisture for evaporation. This will help you to stay cool. In a cold climate, the old saying is "cotton kills". The cell structure of cotton collapses when wet. This destroys its ability to insulate you from the cold.

A good hat and gloves are necessary; you lose up to 50% of your body heat though an uncovered head. Do not forget good footwear, preferably boots with 2 extra pair of heavy socks. The boots should be comfortable and well broken in. The clothing should be subdued in color.

Personally, I avoid goose down. An over dependence of down has resulted in deaths. Down loses most of its insulation properties when it gets wet.

Maps - Get good topographical maps of the area in which you intend to travel.

Photos - Every pack, especially children's, should contain a family photo. If you are separated, having a picture to show others is worth a thousand words. This is particularly important for young children that have trouble communicating. This might mean the difference in ever locating your missing child.

Cash - One hundred dollars in small bills. One-dollar bills are the preferred choice; it might be hard to make change.

Important papers - Do not forget your important papers; for example, insurance, identification, passports and financial, etc.

First aid kit - It should contain a minimum of the following items packed in a waterproof container if possible.

Surgical dressing, approx. 4" x 6"
Band-aids, assorted sizes
Gauze pad, 4", four each
Gauze pad, 2", four each
Adhesive tape, ¾"
First aid book
Vaseline
Triangular bandage
Antiseptic
Anti-diarrhea medicine
Aspirin and acetaminophen
Calamine lotion
Cotton swabs
Ace bandage, 3"

Moleskin for blisters
Sunscreen
Mosquito repellant
Prescription medicine as needed

Get home kits - I keep a get home kit in my car. This is different from a 72-hour kit. This kit is designed to help you get home in an emergency. The contents of this kit will depend on two things: What type of area do you live in, and what are your traveling habits? I spend quite a bit of time out in the country and my kit is designed accordingly. The following is an inventory of the kit that is currently in my car.

2 Military battle dressings
1 Compress and bandage, 2x2, camouflaged
1 Ace bandage, 2 in
2 Cera Lyte 70 oral electrolyte packages
2 Safety pins
3 Instant wash towels
5 Packages Bacitracin antibiotic ointment + pain reliever
3 Packages benzalkonium chloride
3 Bandages
1 Mole skin
2 Knuckle bandages
Aspirin
Tylenol
Claritin
Benadryl
1 3x4 Surgical pad
1 3x4 Telfa pad
1 Triangular bandage

2 Pair surgical gloves

1 Basic Survival Fishing Kit

1 Master blaster fire starter
1 Waterproof match box and matches

1 Military soap dish containing the following:
1 Magnesium fire starter
1 35mm can of cotton balls and petroleum
3 Prepackaged fire starters

2 One gallon plastic bags
1 Flashlight
1 Spoon
6 Misc. size cable ties

2 Emergency ponchos from Emergency Essentials
2 Emergency sleeping bags from Emergency Essentials, space blanket type material
1 8x10 foot tent footprint from REI with tie downs (makes a good emergency shelter)
1 Hundred foot of Mil spec 550 cord

1 Package Chlorine Dioxide tablets by Aquamira
1 Aquamira Water filter

1 Sheath knife
1 Knife sharpener
1 Unbelievable saw

1 Plastic box containing the following:
 Spare flashlight

Compass
2 Broad arrow heads (fishing spears or frog gigs)
Misc. wire

2 Packages Datrex 3600-calorie emergency rations
3 MRE entrees
8 Cliff or granola bars
6 MRE sides
1 Can Vienna sausages
1 Can kippers
30 Butterscotch candies

2 Military 2 quart canteens

Everyone's get home kit will be different; if you work close to home, yours may be much smaller. Your kit may vary depending on the weather.

Chapter 14 - Survival

This chapter will not be a complete primer on survival. That would take several books as well as field training. However, this chapter will give you some basic principles and resources.

Remember the acronym STOP.

S – Stop
T - Think
O - Observe
P – Plan

Stop – Avoid panic, calm down and relax.

Think - Before you do anything, think first. Do not move without considering the consequences.

Observe - Look at your surroundings. What supplies and skills do you have? Is anyone with you, and what are their skills and condition?

Plan - Set your goals. What is your most important need? Is it first aid, water, shelter, fire, food, signaling for help, or communications?

Travel under Survival conditions

If you are lost in a situation where rescuers will be searching for you, stay where you are. Only move if there is immediate danger; for example, a forest fire or flood.

If rescuers are searching for you and you have to move, leave a note indicating your direction of travel and regular trail markers.

Remember some basic rules;

Use your head, try and place yourself in relation to known locations. For instance, in California, if you are lost in the mountains, you know that the populated areas are mainly to your west.

Stay on trails if possible.

In the desert heat, do not travel during the heat of the day.

Do not travel in bad weather. Try to find shelter instead.

Streams or rivers run downhill, usually toward civilization.

Avoid going in circles. Find distant landmarks and travel straight towards them.

If you have a compass, use it to travel in a straight direction that leads you to a known location or landmark. For example, if you are lost in Nevada and you know that you are south of Highway 80, use you compass to travel north.
Determining north without a compass - If you do not have a compass, there are several ways to determine the four points of the compass.

The sun will show you east and west in the early morning and evening. In the northern hemisphere, the path of the sun in the winter is south of the zenith. The zenith is the point directly over your head. In the summer, the path of the sun is almost directly overhead.

During the winter months, the sun is to your south at noon. At this time, shadows will point to the north. If you are in the southern hemisphere the opposite is true.

An ordinary watch with hour and minute hands will help you find true north. In the northern hemisphere, point the hour hand to the sun. The north south line will be found midway between the hour hand and twelve noon. See following diagram.

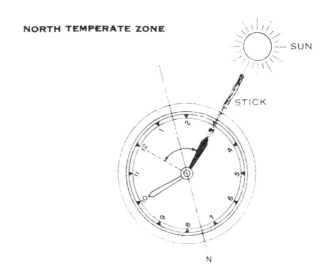

If it is a cloudy day, stand a stick over the center of the watch. Hold it so that the shadow falls along the hour hand. North is halfway between the hour hand and 12 noon. See the following diagram:

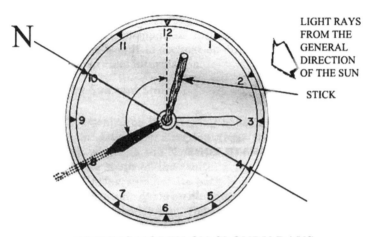

FINDING NORTH ON CLOUDY DAYS

In the northern hemisphere, the North Star is never more than 1 degree of true north. The easiest way to find the North Star is to find the Big Dipper. The following diagram shows what it looks like. Once you find the Big Dipper run a line from the two bottom stars in the direction that you would pour something out of the dipper. This will point right to the North Star.

The Big Dipper.

Stand a 3-foot stick on end and place a small rock where the tip of the shadow falls. Wait ten to fifteen minutes and place a second rock at the point where the tip of the shadow has moved too. Draw a line between the two points. This is an east-west line. Place the toe of your left foot at the first rock and the toe of your right foot at the second rock; you will now be facing north.

If you are ever uncertain which direction the lines runs, remember that anywhere on earth, the first shadow mark will always be west, the second east.

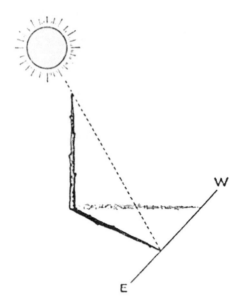

Shadow Method of finding directions

Shadow Method of Finding Directions

Food

Food is normally not a high priority. A healthy human can go three weeks without eating if they have water and shelter.

Edible plants - Native Americans utilized many of the plants in your area for both food and medicinal purposes. Plant knowledge is an excellent method of supplementing your diet in an emergency. Many of the survival books published by the military and other sources are too general. They are designed for worldwide use. If you look in your local bookstores, check with the county agricultural department and local colleges, you should be able to find an edible plant book specific to your area.

Spend a little time talking to older residents; they are often knowledgeable about the local plants and very willing to share information. Never assume that if an animal eats it, you can too. That could be your last mistake.

Universal Edibility Test – The test was developed by the U.S. Military. The test has become controversial and I would only suggest using it in a life or death emergency.

Tasting or swallowing even a small portion of some plants can cause severe discomfort, extreme internal disorders, and even death. Therefore, in a life or death situation, if you have doubts about a plant's edibility, apply the Universal Edibility Test before eating any portion of it. Just complete the following steps:

1. Test only one part of a potential food plant at a time.

2. Separate the plants into their basic components — leaves, stems, roots, buds, and flowers.

3. Smell the food for strong or acidic odors. Remember, smell alone does not indicate a plant is edible or inedible.

4. Do not eat for 8 hours before starting the test.

5. During the 8 hours, abstain from eating. Test for contact poisoning by placing a piece of the plant part you are testing on the inside of your elbow or wrist. Usually 15 minutes is enough time to allow for a reaction.

6. During the test period, take nothing by mouth except purified water and the plant part you are testing.

7. Select a small portion of a single part and prepare it the way you plan to eat it.

8. Before placing the prepared plant part in your mouth, touch a small portion (a pinch) to the outer surface of your lip to test for burning or itching.

9. If after 3 minutes there is no reaction on your lip, place the plant part on your tongue, holding it there for 15 minutes.

10. If there is no reaction, thoroughly chew a pinch and hold it in your mouth for 15 minutes. **Do not swallow.**

11. If no burning, itching, numbing, stinging, or other irritation occurs during the 15 minutes, swallow the food.

12. Wait 8 hours. If any ill effects occur during this period, induce vomiting and drink a lot of water.

13. If no ill effects occur, eat 1/4 cup of the same plant part prepared the same way. Wait another 8 hours. If no ill effects occur, the plant part as prepared is safe for eating.

Beware -Test all parts of the plant for edibility, as some plants have both edible and inedible parts. Do not assume that a part that proved edible when cooked is also edible when raw. Test the part raw to ensure edibility before eating raw. The same part or plant may produce varying reactions in different individuals.

Before testing a plant for edibility, make sure there are enough plants to make the testing worth your time and effort. Each part of a plant (roots, leaves, flowers, and so on) requires more than 24 hours to test. Do not waste time testing a plant that is not relatively abundant in the area.

Eating large portions of plant food on an empty stomach may cause diarrhea, nausea, or cramps. Two good examples of this are familiar foods such as green apples and wild onions. Even after testing plant food and finding it safe, eat it in moderation.

You can see from the steps and time involved in testing for edibility just how important it is to be able to identify edible plants. Get a good book that covers your area.

There are some general rules to help you avoid potentially poisonous plants. Stay away from any wild or unknown plants that have:

- Milky or discolored sap.
- Beans, bulbs, or seeds inside pods.
- Bitter or soapy taste.
- Spines, fine hairs, or thorns.
- Dill, carrot, parsnip, or parsley like foliage.
- Almond scent in woody parts and leaves.
- Grain heads with pink, purplish, or black spurs.
- Three-leaved growth pattern.

If you use the above criteria to eliminate plants when using the Universal Edibility Test, you will avoid some edible plants by mistake. More importantly, the criteria will help you avoid plants that are potentially toxic to eat or touch.

Fish traps - In surf, shallow areas, and small streams it is possible to trap fish. Before you decide to set up a fish trap, understand that it requires effort. Are the calories you expend making the trap worth the calories you gain from the fish? If you are going to be in an area several days, a fish trap is probably a good idea.

A fish trap consists of a series of sticks or rocks laid in a pattern that traps the fish. You build a corral with a funnel leading into it. When you have fish trapped, block the entrance and either club the fish, or catch them with your hands. If you get a large amount of fish in the trap, you can leave them caged up until you are ready to eat them.

Remember that in large streams and lakes, fish migrate into the shallows along the banks in the early morning and evening to feed.

The trap can block a small stream completely. The tidal fish trap is the best choice for these conditions. Wade in upstream and herd the fish into the trap.

In the ocean schools of fish approach the shore at high tide and travel along the shore. Set your traps when the tide is out. The v-type trap works well under these conditions. It will trap fish as the tide goes out.

Use natural rock pools, sand banks or other natural features whenever possible. See the following diagram for an example of a fish trap.

Tidal fish *trap*.

V-type *trap*.

Animal traps

Figure four trap-These can be used with a deadfall. A deadfall can be made for either large or small game. If you decide to try for large game, consider the amount of work versus your chances of success. See the diagram that follows.

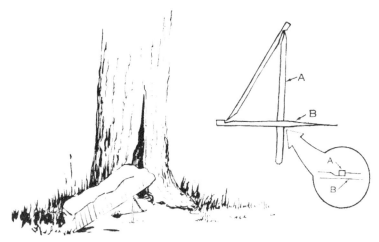

A simple deadfall using a figure 4 trigger.

Fixed snares - This type of snare is particularly good for rabbits and other small game. Set a loop with a slipknot across a game trail and anchor it to a log, tree, bush, notched or forked stick. When the animal sticks its head through the loop, it will set off the trap strangling itself.

Either wire or a light cord can be used; just be sure that the loop slides free. You can bait the trap with fresh dandelions, or other leafy greens. See the sample in following diagram.

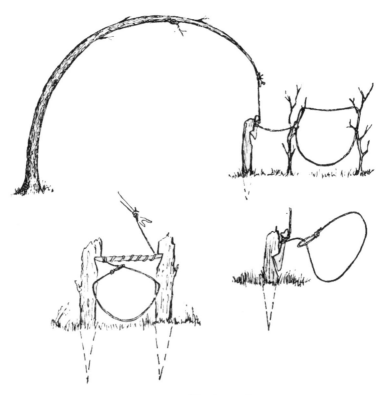

Fixed snare.[7]

Hanging snares - Make a loop with a slipknot and attach it to a small sapling. The sapling needs to be strong enough to lift a small animal. Bend the sapling over and anchor the loop to a trigger as shown in the following diagram. Make sure the loop is big enough to fit over the animals head.

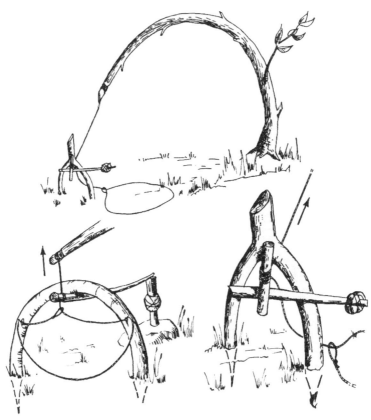

Hanging snares.

Trapping hints - All mammals and birds are edible. Rabbits, squirrels, rats, and mice are relatively easy to trap. They are creatures of habit and normally confine their activities to small areas. Just find a hole or trail and set your trap. If you can determine what type of food they have been eating, use it to bait your trap.

If you have an area for butchering animals, this is a good place to set snares. At night set snares in game trails containing droppings or fresh tracks.

If you are near water, place a minnow on a fishhook and place it near the shore. There is a chance a bird will grab it. Do not forget to tie a line to the fishhook and anchor it to a bush or tree. With the right type of bait, you can catch ducks.

Fire starting - In an emergency if the sun is shining, you can use a lens from your glasses, binoculars, or a camera lens to concentrate sunlight and start a fire.

Figure 135. Sun and glass.

In Chapter 13 on 72-hour kits, there is a list of fire starters and suggestion on what you should carry. I always carry a fire starting (magnesium, flint, and steel) device on my key ring. It does not matter what methods you choose to use as long as you practice the techniques and have the necessary materials available.

However, learning primitive methods that are not dependent on modern manufactured fire starters is a good idea.

Fire starting tips - Do not build your fire too large. Several small fires in a circle around you will provide more warmth.

Use kindling to start your fire. This can consist of small dry twigs, grasses, pine knots, leaves, pine needles, and any other dry flammable material you can find. If it is wet weather, break open dead decayed trees and stumps and look for punk wood. Look for pine resin; you will sometimes find it around the remains of pine knots in downed trees. Pine resin or sap will burn when wet. Scrape it into a powder to make it easier to light. Cut your dry wood into shaving before you light it.

Once your fire is started, add wood slowly. The lower branches of trees are often dry even during a rain.

Bank your fire at night. Use a green log or the butt of a decayed punky log to keep your fire burning slowly. Cover the embers with a thin layer of ash or dirt.

Done right you should be able to keep your fire going all night.

Avoid starting grass or forest fires. Carefully pick the area in which you will start your fire. Make sure it is clear of dried grass brush and other flammables. In the forests scrape away the duff and make sure you have bare soil. Duff consists of shed vegetative parts, such as leaves, branches, bark, and stems, existing in various stages of decomposition above the soil surface. Duff will burn and fire can travel through it undetected for some distance.

When you leave make sure your fire is extinguished. Wet the embers down or cover them with a thick layer of dirt. Never leave a fire unattended.

A windbreak like in the following diagram will help protect your fire from the wind. It can also serve as a heat reflector as well as protect you from the wind.

Windbreak or reflector.

Clothing - I live in a mountainous area. Every year I hear of deaths caused from hypothermia that would have not occurred if the right clothing had been worn. This is a lack of planning.

In cold weather, wear multi layers. They should start with a good base of long underwear. Avoid the cheap cotton long underwear sold in the discount houses. Fleece, polypropylene, or wool retain most of their insulating properties when wet. They are the best choices. Cotton loses its ability to insulate when wet.

You should have a good intermediate layer. This should consist of a shirt and pair of pants made from fleece, polypropylene, or wool. You should never wear blue jeans. As an option (one I prefer) you may want to add a vest.

The outer layer should be water/wind resistant, preferably with a hood. Wear a good pair of gloves and do not forget to cover your head. In some circumstances you can lose up to 50% of your body heat through your uncovered head.

Keep your feet warm with two pairs of socks and a good pair of boots. This is a minimum for cold weather.

The reason that layering works is that it gives you the ability to control your body heat. You can keep from sweating or getting cold by adjusting the layers. Stay warm, do not have to work to get warm and start to sweat.

In hot desert climates, protect yourself against excessive evaporation and sunburn by staying properly dressed. Wear long pants and long sleeved shirts. Your clothes should be worn loosely.

Wear a hat and have a cloth neckpiece covering the back of your neck. Remember it gets cold in the desert at night. Do not throwaway extra clothing during the heat of the day.

If you do not have boots, make puttees if you have extra cloth. Take two 3 or 4 inch wide strips of cloth about four feet long. Wrap them spiral fashioned from the top of your shoes upward over your pants. This will keep out most of the sand and insects.

Shelter - If you get too hot or too cold, you can die. Finding good shelter can help you keep a normal body temperature. Shelter will help keep you dry and out of the wind or sun.

If you have a vehicle with you, you can utilize it for shade or shelter. If you are in the desert, the vehicle may become too hot during the day. You may be better off sitting next to it for shade. Remove a car seat and use it to keep you off the hot ground.

In hot weather, ground level is the hottest place. Try to get below or above the ground level. Dig a hole; build a seat or a hammock.

If there is natural shelter such as a cave, fallen tree, hollow log, rock overhang, or brush, take advantage of it, but watch for animals.

In cold weather, you can add insulation by stuffing your clothes with dry grass, leaves or carpeting from your vehicle. Just be sure it is dry and insect free.

If it is cold use leaves, grass, tree boughs, pine needles, or anything else you can think of to insulate you from the ground. The rule for staying warm is to put 2/3 of the insulation underneath you and 1/3 on top.

Low areas like valleys can be colder in the winter. It can be several degrees warmer if you just walk uphill a short distance.

Share warmth; do not be afraid to huddle together.

Plastic bags, plastic sheets, or space blankets do not breathe. If you wrap them about you too tightly, they will trap moisture. Do not wrap plastic about your head as it can suffocate you.

If you are forced to sit in a confined area like a snow cave, do isometric exercises (tension exercises). This will increase your body heat.

See the shelter section of Chapter 13 on 72-hour kits for more information.

In the following diagram, a shelter from the wind, an insulated bed of leaves, pine needles, boughs, a fire, and a reflector will help you stay alive.

Method of pitching lean-to.

Tree well shelter - A simple quick shelter that is easy to make. First, find a thick sturdy tree in deep snow. Conifers work best.

Then dig a hole in the snow near the tree's base using whatever you have available including your hands. Try to make the hole at least 4 feet deep. The lower branches of the tree should form an overhead shelter when you are finished.

If the snow is not deep enough to form a roof, gather up branches and use them to make a roof. Pile snow on top of the branches to complete the roof.

Use other branches, pine needles and leaves to create insulation in your shelter by lining the bottom, and perhaps the sides, with them.

When you lay down for the night, curl up in a fetal position, to preserve warmth. Working hard creating the tree well shelter should have warmed you up. You will take that warmth with you into your shelter. Be careful during the construction to not sweat or get wet.

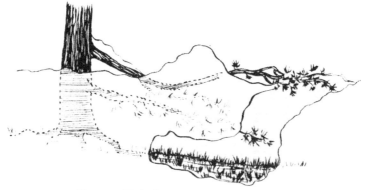

Tree well shelter

Snow Caves - Find a site on the lee side of a hill. Snow caves can be created quickly by digging into a snow bank or drift. Dig a compartment so that it is at least large enough inside for you to sit upright. Place your pack or a block of snow in front of the entrance hole. Use evergreen boughs or other natural materials to insulate yourself from the ground.

You can use a candle or build a very small fire in a snow cave. This requires a vent hole for adequate ventilation. If you have a problem with dripping water, your fire may be too large. Smoothing the inside of the roof helps to stop dripping.

If you think people will be out looking for you, make the site as visible as possible from the ground and the air. Place clothing, sticks or stomp an unusual pattern in the snow. When you are inside the cave your ability to hear what is happening outside will be reduced to almost nothing.

A properly made snow cave can be 0 °C (32 °F) or warmer inside, even when outside temperatures are −40 °C (−40 °F).

Remember to stay dry while building your cave.

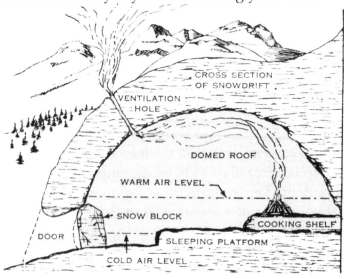

Snow caves.

Hypothermia - is when your body loses heat faster than it can produce heat, resulting in a dangerously low body temperature. The normal body temperature is approximately 98.6° F (37° C). Hypothermia occurs as your body temperature passes below 95° F (35° C).

When your body temperature drops, your heart, nervous system and other organs will not function correctly. Untreated, hypothermia eventually leads to the complete failure of your heart and respiratory system and to death.

Hypothermia is often caused by exposure to cold weather or immersion in a cold body of water. The only treatment is to warm the body back to a normal temperature. Just covering the victim in warm bedding will not work. You need to add heat to the victim, by shared body heat, hot drinks, or another source of heat.

A canteen of hot water or a few hot rocks wrapped in a cloth placed with the victim can make the difference between life and death.

Heat exhaustion - can range in severity from mild heat cramps to potentially life-threatening heatstroke. Symptoms of heat exhaustion often start suddenly, sometimes after excessive exercise, heavy sweating, and not drinking enough liquids. Symptoms resemble those of shock and may include:
- Feeling faint or dizzy
- Nausea
- Heavy sweating
- Rapid, weak heartbeat

- Low blood pressure
- Cool, moist, pale skin
- Low-grade fever
- Heat cramps
- Headache
- Fatigue
- Dark-colored urine

If you suspect heat exhaustion:

Move the person out of the sun and into the coolest shady space you have available.
Lay the person down and elevate the legs and feet slightly.
Loosen or remove the person's clothing.
Have the person drink any available cool water or non-alcoholic caffeine-free drink.
Cool the person by spraying or sponging them with cool water and fanning.
Monitor the person carefully. Heat exhaustion can quickly become heatstroke.
If fever greater than 102° F (38.9° C), fainting, confusion or seizures occur, call 911 or emergency medical help at your earliest opportunity.

Heatstroke - constitutes the most severe of the heat-related problems. Working or exercising in hot environments such as desert areas combined with inadequate fluid intake can result in heatstroke.

Young children, older adults, and people who are overweight are at high risk of heatstroke. Dehydration and exposure to the sun are often the primary cause in survival situations.

Heatstroke is potentially life threatening because the body's normal mechanisms for dealing with heat stress are inadequate. The main sign of heatstroke is an elevated body temperature, normally greater than 104° F (40° C). It can include mental changes ranging from personality changes to confusion and coma. The skin may be hot and dry.

- Other signs and symptoms may include:
- Rapid heartbeat
- Rapid and shallow breathing
- Elevated or lowered blood pressure
- Cessation of sweating
- Irritability, confusion, or unconsciousness
- Feeling dizzy or lightheaded
- Headache
- Nausea
- Fainting, may be the first sign in older adults

If you suspect heatstroke:
Move the person out of the sun and into the coolest shady space you have available.
Cool the victim by covering them with damp clothing or by spraying with cool water. Place damp clothes on the back of neck and forehead. Do whatever you can to cool the person down. Fan them.

Have the person drink any available cool water or non-alcoholic caffeine-free drink, if they are able.
Call 911 or emergency medical help at your earliest opportunity.

Dehydration - occurs when you lose more body fluid than you take in. Your body lacks the water and other fluids to carry out its normal functions. If you do not replace the lost fluids, you will get dehydrated.

Causes of dehydration include intense diarrhea, vomiting, fever, or excessive sweating. Not drinking adequate amounts of water during hot weather or exercise may cause dehydration. Only 5% dehydration can cause confusion that can result in you making bad decisions. Healthy adults can become dehydrated, but young children, older adults, and people with chronic illnesses are most at risk. Exposure to cold, heat, intense activity, and high altitudes, illness, or burns will cause your body to use more water than normal.

Mild to moderate dehydration can usually be reversed by drinking more liquids. However, severe dehydration needs immediate medical treatment. The safest approach is prevention. Monitor your fluid loss during hot weather or exercise, and drink enough liquids to replace what you lose.

The common signs for dehydration are

- Dark urine/strong odor
- Urine level low
- Skin stays up when pinched

- Dark sunken eyes
- Extremely tired
- Confusion and emotional problems
- When fingernails are squeezed, color delays returning.

Water - Your body is approximately 60 % water. You can survive weeks without food, but only about 1 – 7 days without water depending on the temperature. Clean drinkable water is often your most important survival consideration.

Ration sweat not water; in other words try to keep from losing body fluids through sweat.

When possible boil or filter your water to purify it. If you do not have the means to purify water you may have to make the decision to drink it anyway. Look around for the cleanest running water you can find. If it is muddy, try filtering it through a piece of cloth. You may not get sick   However, it may be better to take the risk of drinking polluted water than die of dehydration.

Do not drink salt water, alcohol, blood, or urine. Your body uses more water processing these liquids than you gain.

Do not eat snow or ice without melting. It will lower the core temperature of your body. Melting ice provides much more water than melting snow.

In dry areas, you may find water by digging for water inside the bends of dry rivers and washes, or near green vegetation. Dig down about two feet and if you find damp ground wait for water to collect. You can get it out of the hole with a straw or cloth to soak up water. Dig your hole at night or in the shade to avoid overheating.

Wipe down metal, rocks or leaves at sunrise to gather dew.

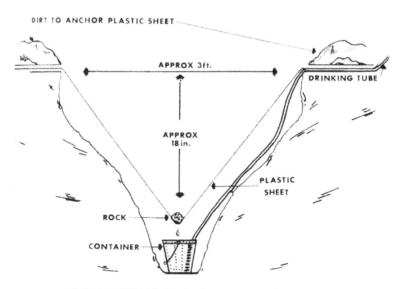

CROSS SECTION OF SURVIVAL STILL. HEAT FROM SUN VAPORIZES GROUND WATER. THEN THIS VAPOR CONDENSES UNDER PLASTIC, TRICKLES DOWN, DROPS INTO CONTAINER.

Water "still".

If you have a sheet of plastic, you can dig a solar still like in the previous diagram. You can place non-toxic green leaves and grass in the hole to make it more efficient. Just remember that stills are not fast and do not provide a large amount of water. Nevertheless, it all helps.

Chapter 15 - Improvised Equipment and Supplies

This chapter will cover some simple, old-fashioned recipes and tricks as well as improvised equipment that will help make your life more comfortable.

Toothpaste - Equal amounts of salt and powdered sage leaves makes good toothpaste.

Cures for Insects and Rodents - If skunks get into your garage or under your house, throw a handful of mothballs into the area. They hate the smell and will leave almost immediately.

A piece of cloth well saturated with cayenne pepper can be used to plug rat and mouse holes.

Cockroaches and ants have a dislike of cayenne pepper.

Sassafras oil drops will drive ants away. Put a few drops on the shelves.

Branches of elder bush hung in the house will help keep flies away. They do not like the odor.

Eucalyptus tree leaves placed in a closet will keep moths away.

African daisies (pyrethrum), dried and powdered, are an excellent insecticide. They are the basis for many modern commercial insect repellants.

Insects do not like bay leaves. Make little cloth a bag of bay leaves and place in cupboards or drawers to keep insects away. Bay trees grow wild in many areas of the country.

Weed killer - The recipe for natural weed killer is simple. Combine a gallon of white vinegar with 1 cup of salt and a tablespoon of liquid dish soap. Stir it all up and pour or spray on the weeds.

Use this and you won't have weeds for 3 to 6 months. Don't get this natural weed killer on any plants you want to keep because it's actually a soil sterilizer. It kills all types of plants.

Insect bites - Plantains can be crushed and used to reduce itching from mosquito bites or relieve pain from bee stings.

Matches and Fires - Waterproof your matches by melting wax into a small bowl and dipping about half the length of a match into the melted wax.

Save the grease and fat from your cooking. Pour it over shredded paper or tinder to make improvised fire starters.

Kindling can be dried out in your solar oven.

Improvised equipment – This section will show you some simple ways to improvise common items.

Keeping your food cool helps to prolong its edible life and makes many things more appetizing. Your useless electric refrigerator is well insulated and can be effective as a cooler. Picnic coolers are very useful. They can be set in the ground and covered up, but be wary of animals.

An iceless refrigerator is simple and cheap to make and will keep food cool. At one time, these were common all over the Southwestern United States. They work best in a dry climate.

Instructions for Making an Iceless Refrigerator

Make a wooden or PVC pipe frame approximately 48 - 58 inches high by 12 -18 inches wide and deep. Cover the outside with a wire screen or hardware cloth. The covering should be something that preferably will not rust and will keep insects out. The top should be screen and the bottom solid. Make a door for one side and hang it on hinges or leather straps. Use a hook or a button for a latch.

The inside shelves can be fixed or adjustable and should be made of a frame covered with wire fencing such as chicken wire. This will allow airflow.
Paint the wooden frames with whatever type of paint you have available. If you do not have paint, use linseed or cooking oil and give it time to dry.

Next, make a cover of burlap, flannel, or any heavy, coarse, water-absorbent cloth. Put the smooth side of the material on the outside. Attach the cover around the top of the frame with nails, buttons, etc. Leave a flap for the door to open.

Place a 2-4 inch deep pan on top of the frame. Sew strips of cloth to the top of the cover and extend them into the top pan. They will act as a wick to draw water into the cover.

Place a second pan under the bottom of the frame and make sure the bottom of the cover reaches into it.

Stand the refrigerator in a shady place where there is good airflow. The refrigerator works when you fill both pans with water. The water wicks into the cover and saturates it. The airflow causes an evaporative action similar to a swamp cooler. You can jump-start the refrigerator by first pouring water over the cover. The faster the rate of evaporation gets, the cooler the temperature.

A collapsible variation of this refrigerator was often used at campsites. It was hung from a tree branch out of the reach of animals.

Iceless Refrigerator

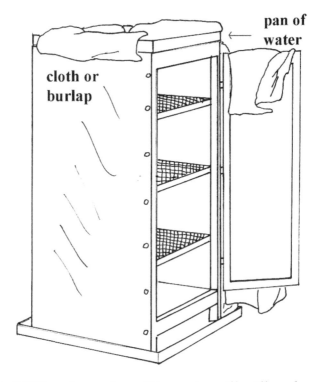

Grain Mills - A good quality grain mill will make your life much easier. There is currently a disagreement between proponents of stone ground and metal-jawed mills. Some feel that the heat generated by the stones causes a loss of nutrition.

I feel that either is a good choice. Before you purchase a mill, try it. You might be surprised at the amount of strength it takes to use it. I knew someone who powered theirs with a bicycle. If you intend to do this get the parts now and perfect it before you need it.

Here is a diagram of an alternate method of cracking your grain. This is a method of last resort; it is slow, hard work.

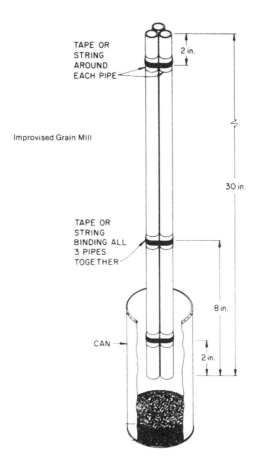

Improvised Grain Mill

Place 1 to 2 inches of grains in the can and use the pipes to crack the grain. This will not make fine grain flour, but just cracking it will save you a lot of cooking and soaking time.

Solar ovens

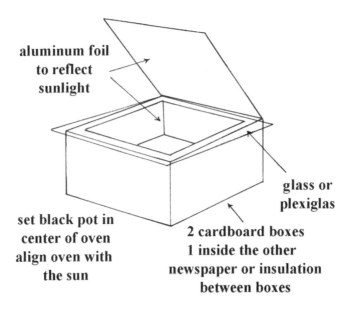

The outside dimensions of a solar oven should be approximately 24 inches square. The inner box should be approximately 20 inches square. The space in between the two boxes should be filled with insulation. This can consist of newspaper or cardboard. Line the inside and the top reflector with aluminum foil and put a lid of glass or Plexiglas on top and you are ready to cook.

Rocket stove

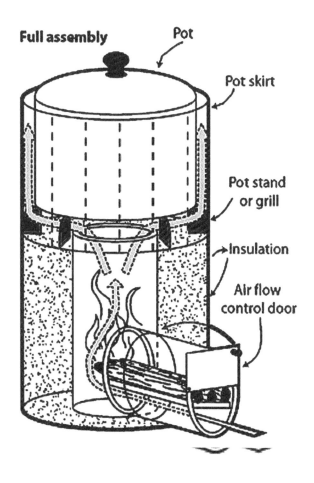

A five-gallon bucket works well for the outer container.

Winiarski / Aprovecho rocket stove

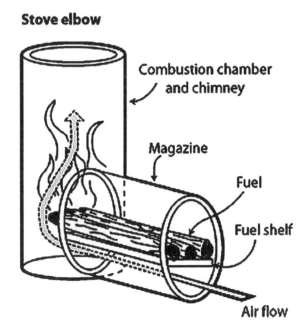

A four-inch pipe works well for the combustion chamber and chimney. I have seen these made using stovepipe, but I prefer something a little heavier. Dirt or crushed lava rock makes a good insulation. The fuel shelf can be made out of an old tin can. Using the same principal, you can improvise these stoves out of brick or tile.

Rocket stoves are quite efficient and are good to have if firewood is in short supply. They will make a good hot fire out of twigs and other small combustibles.

TOOLS - Do not forget the old pioneer standby - a good axe. They cleared fields and built houses with not much more. If you are planning on burning wood, have the tools to cut it. Do not depend on your chain saw. The same principle applies if you are planning a garden: shovels, spades, forks, rakes, etc. You cannot think of everything: just do your best. In addition, remember, if you have anything extra, you can always trade.

Chapter 16 - Weapons of Mass Destruction

A weapon of mass destruction (WMD) is any weapon that can kill or harm a large number of humans and cause great damage to man-made structures.

Nuclear explosion - First, understand that nuclear warfare is survivable with a little luck, skill, and training. I believe that eventually there will be a nuclear incident in the United States. Either terrorists or a foreign power will cause the incident. It will be either a dirty bomb or a nuclear explosion.

There are several good sources of information on how to survive a nuclear incident. Nuclear War Survival Skills, by Cresson H. Kearny is, in my opinion, the best book ever written on this subject. Get a copy of this book, read it, and learn how to protect yourself. Some new information is starting to appear from the Federal Government. A publication that is available though FEMA or on the web called "Planning Guidance to Response to a Nuclear Detonation, Second Addition" is quite good. It is based on a terrorist attack with a small nuclear (10KT or less) explosion.

If you are within the blast radius, expect damage for small devices to be in accordance with the following table:

Approximate distances for zones with varying yield nuclear explosions.

10 KT Explosion
- The Severe Damage Zone will extend to ½ mile
- The Moderate Damage Zone will be from ½ mile to 1 mile
- The Light Damage Zone will extend from 1 mile to 3 miles

1 KT Explosion
- The Severe Damage Zone will extend to ¼ mile
- The Moderate Damage Zone will be from ¼ mile to ½ mile
- The Light Damage Zone will extend from ½ mile to 2 miles

0.1 KT Explosion
- The Severe Damage Zone will extend to 200 yards
- The Moderate Damage Zone will be from 200 yards to ¼ mile
- The Light Damage Zone will extend from ¼ mile to 1

Light Damage (LD) Zone:

Damage is caused by shocks, similar to those produced by a thunderclap or a sonic boom, but with much more force. Although some windows may be broken over 10 miles away, the injury associated with flying glass will generally occur at overpressures above 0.5 psi. This damage may correspond to a distance of about 3 miles from ground zero for a 10 KT nuclear explosion. The damage in this area will be highly variable as shock waves rebound multiple times off buildings, the terrain, and even the atmosphere.

As a responder moves inward, windows and doors will be blown in and gutters, window shutters, roofs, and lightly constructed buildings will have increasing damage. Litter and rubble will increase moving towards ground zero and there will be increasing numbers of stalled and crashed automobiles that will make emergency vehicle passage difficult.

Blast overpressures that characterize the LD zone are calculated to be about 0.5 psi at the outer boundary and 2-3 psi at the inner boundary. More significantly, structural damage to buildings will indicate entry into the moderate damage zone.

Moderate Damage (MD) Zone:

Responders may expect they are transitioning into the MD zone when building damage becomes substantial. This damage may correspond to a distance of about one mile from ground zero for a 10 KT nuclear explosion.

Observations in the MD zone include significant structural damage, blown out building interiors, blown down utility lines, overturned automobiles, caved roofs, some collapsed buildings, and fires. Some telephone poles and street light poles will be blown over. In the MD zone, sturdier buildings (e.g., reinforced concrete) will remain standing, lighter commercial and multi-unit residential buildings may be fallen or structurally unstable, and many wood frame houses will be destroyed.

Substantial rubble and crashed and overturned vehicles in streets are expected, making evacuation and passage of rescue vehicles difficult or impossible without street clearing. Moving towards ground zero in the MD zone, rubble will completely block streets and require heavy equipment to clear.

Within the MD zone, broken water, gas, electrical, and communication lines are expected and fires will be encountered.

Many casualties in the MD zone will survive, and these survivors, in comparison to survivors in other zones, will benefit most from urgent medical care.

A number of hazards should be expected in the MD zone, including elevated radiation levels, potentially live downed power lines, ruptured gas lines, unstable structures, sharp metal objects and broken glass, ruptured vehicle fuel tanks, and other hazards.

Visibility in much of the MD zone may be limited for an hour or more after the explosion because of dust raised by the shock wave and from collapsed buildings. Smoke from fires will also obscure visibility.

Blast overpressures that characterize the MD zone are an outer boundary of about 2–3 psi and inner boundary of about 5–8 psi. When most buildings are severely damaged or collapsed, responders have encountered the severe damage zone.

Severe Damage (SD) Zone:

Few, if any, buildings are expected to be structurally sound or even standing in the SD zone, and very few people would survive. However, some people protected within stable structures (e.g., subterranean parking garages or subway tunnels) at the time of the explosion may survive the initial blast.

Very high radiation levels and other hazards are expected in the SD zone, significantly increasing risks to survivors and responders. Responders should enter this zone with great caution, only to rescue known survivors.

Rubble in streets is estimated to be impassable in the SD zone making timely response impractical. Approaching ground zero, all buildings will be rubble, and rubble may be 30 feet deep or more.

The SD zone may have a radius about a 0.5 mile for a 10 KT detonation. Blast overpressure that characterizes the SD zone is 5-8 psi and greater.

The following description of the scope and size of damaged areas has been taken from "Planning Guidance to Response to a Nuclear Detonation, Second Addition"

Thermal damage has not been mentioned in the Zones of destruction. A nuclear blast will cause a large burst of thermal energy. This will start fires and burn exposed individuals. The majority of the fire will occur in the Moderate Damaged Zone. The wind will be too high in the Severe Damage Zone. There may be some fires in the Light Damage Zone.

If you are within one of these three zones, the only advice I can give you is to get out if you can or find the best shelter that is available.

There is always the chance of a larger nuclear device in the one-megaton or more range. These will create the same zones of destruction only larger. The Light Damage Zone would extend to approximately 8 miles with some windows broken beyond that.

Since the majority of us would not live within blast areas, we have to be concerned with protection from fallout.

Protection from Fallout:

People who are not threatened by blast and fire, but who need protection against fallout, need to consider three factors: distance, mass, and time.

The more distance between you and the fallout particles, the less radiation you will receive. In addition, you need a mass of heavy, dense materials between you and the fallout particles. Materials like concrete, bricks, and earth absorb many of the gamma rays. Over time radioactivity in fallout loses its strength.
The decay of fallout radiation is expressed by the "seven-ten" rule. Simply stated, this means that for every sevenfold increase in time after detonation, there is a tenfold decrease in the radiation rate. For example, if the radiation intensity one hour after detonation is 1,000 Roentgens (R)* per hour, after seven hours it will have decreased to one-tenth as much—or 100 R per hour. After the next sevenfold passage time (49 hours or approximately two days), the radiation level will have decreased to one-hundredth of the original rate, or be about 10 R per hour.

The following diagram shows the decreased dose at 49 hours. After about a two-week period, the level of radiation would be at one-thousandth of the level at one hour after detonation, or 1 R per hour.
Radiation exposure is measured in Roentgens (R).

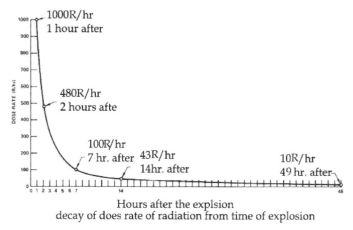

Hours after the explsion
decay of does rate of radiation from time of explosion

One way to protect yourself from fallout is by staying in a fallout shelter. As indicated above, the first few days after an attack would be the most dangerous time. Even starting with a high, close to ground zero, fallout exposure rate of 1000 R/hr, well within two weeks after an attack, the occupants of most shelters could be spending some time outside performing essential chores or attempting to acquire additional supplies, or just evacuating further away. After two weeks, most people will be able to stop using the shelter altogether.

The following diagram provides a summary of the radiation exposure reduction factors as a result of building type and location within the building.

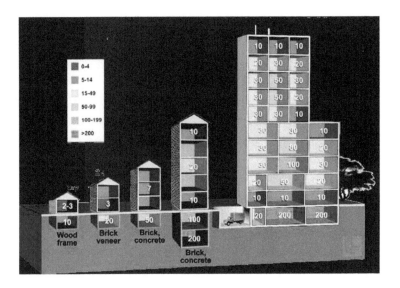

Building as shielding – Numbers represent a dose reduction factor. A dose reduction factor of 10 indicates that a person in that area would receive 1/10th of the dose of a person in the open. A dose reduction factor of 200 indicates that a person in that area would receive 1/200th of the dose of a person out in the open.

For beta and gamma radiation. 1 rad of exposure = 1 rem of dose, and can also be expressed as roentgens.

Fallout effects are potentially avoidable unlike initial effects. From ground zero, to 10 to 20 miles out, unsheltered people could receive acute and even lethal radiation doses. The lethal dose for untreated patients is approximately 400 rads. Medical care increases one's chances of survival up to a dose of 600 rads. Even with medical care, many victims that receive radiation doses fewer than 600 rads would not be expected to survive. The time from exposure to death for these victims ranges from several weeks to a few months. People, who are subjected to acute doses above 200 rad, will likely be unable to perform their jobs adequately and be at risk of becoming casualties themselves. Below the range of acute effects, the risk of cancer is increased over a person's lifetime.

Potassium Iodide K1 will help resist the radioactive iodine that has an affinity for the thyroid gland in your neck. People who do not have protection may be prime candidates for cancer in 10 years. Potassium Iodide only protects you against radioactive iodine. You have no way of knowing if radioactive iodine has been used in a radiological dispersion device. Potassium iodide is harmful to some people and FEMA recommends against taking it without instructions from government authorities.

If you are within the fallout pattern and decide to take potassium iodide, start as soon as possible after a blast. Normal iodine used for medicine on cuts & scratches is POISON if taken internally and would **NOT** protect the thyroid

Radiological Dispersion Device -There is no way of knowing how much warning time there will be before an attack by terrorists using a Radiological Dispersion Device (RDD), so being prepared in advance and knowing what to do is important. To prepare for an RDD event, you should do the following:

Find out from officials if any public buildings in your community have been designated as fallout shelters. If none have been designated, make your own list of potential shelters near your home, workplace, and school. These places would include basements or the windowless center area of middle floors in high-rise buildings, as well as subways and tunnels.
If you live in an apartment building or high-rise, talk to the manager about the safest place in the building for sheltering and about providing for building occupants until it is safe to go out.

During periods of increased threat, increase your disaster supplies to be adequate for up to two weeks (this is a government recommendation, I recommend a lot more). Taking shelter during an RDD event is absolutely necessary. There are two kinds of shelters - blast and fallout. The following describes the two kinds of shelters:

Blast shelters are specifically constructed to offer some protection against blast pressure, initial radiation, heat, and fire. However, even a blast shelter cannot withstand a direct hit from a nuclear explosion.

Fallout shelters do not need to be specially constructed for protecting against fallout. They can be any protected space, provided that the walls and roof are thick and dense enough to absorb the radiation given off by fallout particles.

While the explosive blast will be immediately obvious, the presence of radiation will not be known until trained personnel with specialized equipment are on the scene. Whether you are indoors or outdoors, home or at work, be extra cautious. It would be safer to assume radiological contamination has occurred — particularly in an urban setting or near other likely terrorist targets — and take the proper precautions. As with any radiation, you want to avoid or limit exposure. This is particularly true of inhaling radioactive dust that results from the explosion. As you seek shelter from any location (indoors or outdoors) and there is visual dust or other contaminants in the air, breathe though the cloth of your shirt or coat to limit your exposure. Even if you manage to avoid breathing radioactive dust, your proximity to the radioactive particles may still result in some radiation exposure.

If the explosion or radiological release occurs inside, get out immediately and seek safe shelter. Otherwise, if you are:

Outdoors	Indoors
Seek shelter indoors immediately in the nearest undamaged building. If appropriate shelter is not available, move as rapidly as is safe upwind and away	If you have time, turn off ventilation and heating systems, close windows, vents, fireplace dampers, exhaust fans, and clothes dryer vents. Retrieve your disaster supplies kit and a battery-powered radio and take them to your shelter

from the location of the explosive blast. Then, seek appropriate shelter as soon as possible. Listen for official instructions and follow directions.	room. Seek shelter immediately, preferably underground or in an interior room of a building, placing as much distance and dense shielding as possible between you and the outdoors where the radioactive material may be. Seal windows and external doors that do not fit snugly with duct tape to reduce infiltration of radioactive particles. Plastic sheeting will not provide shielding neither from radioactivity nor from blast effects of a nearby explosion. Listen for official instructions and follow directions.

After finding safe shelter, those who may have been exposed to radioactive material should decontaminate themselves. To do this, remove and bag your clothing (and isolate the bag away from you and others), and shower thoroughly with soap and water. Seek medical attention after officials indicate it is safe to leave shelter. Contamination from an RDD event could affect a wide area, depending on the amount of conventional explosives used, the quantity and type of radioactive material released, and meteorological conditions. Thus, radiation dissipation rates vary, but radiation from an RDD will likely take longer to dissipate due to a potentially larger localized concentration of radioactive material.

Follow these additional guidelines after an RDD event:

Continue listening to your radio or watch the television for instructions from local officials, whether you have evacuated or sheltered-in-place.
Do not return to or visit an RDD incident location for any reason.

EMP attacks are generated when a nuclear weapon is detonated at altitudes a few dozen miles or higher above the Earth's surface. The explosion of even a small warhead would produce a set of electromagnetic pulses that interact with the Earth's atmosphere and the Earth's magnetic field.

"These electromagnetic pulses propagate from the burst point of the nuclear weapon to the line of sight on the Earth's horizon, potentially covering a vast geographic region; doing so simultaneously, moreover, at the speed of light," said Dr. Lowell Wood, acting chairman of the commission appointed by Congress to study the threat. "For example, a nuclear weapon detonated at an altitude of 400 kilometers (248 miles) over the central United States would cover, with its primary electromagnetic pulse, the entire continent of the United States and parts of Canada and Mexico."

"The electromagnetic field pulses produced by weapons designed and deployed with the intent to produce EMP have a high likelihood of damaging electrical power systems, electronics and information systems upon which any reasonably advanced society, most specifically including our own, depend vitally," Wood said. "Their effects on systems and infrastructures dependent on electricity and electronics could be sufficiently ruinous as to qualify as catastrophic to the American nation."

No one seems to know exactly what would survive. Cars with points should survive; newer cars with electronic ignition probably would not.

Unprotected computers may not survive. Small electronics can be protected by a Faraday cage, which is an enclosed ungrounded or grounded metal container. The contents must be insulated from touching the metal sides of the container. A microwave oven is a good example of a Faraday cage. The microwave oven does not have to be in working condition. An old microwave with the cord cut off should protect small electronic items from EMP.

I have not attempted to explain EMP and Faraday cages in detail. For detailed information on grounding large systems, get a copy of Army Technical Manual 5-690.

The following is what FEMA says about EMP.

What is Electromagnetic Pulse?
An additional effect that can be created by a nuclear detonation is called electromagnetic pulse, or EMP. A nuclear weapon exploding just above the earth's atmosphere could damage electrical and electronic equipment for thousands of miles. (EMP has no direct effect on living things.)

   EMP is electrical in nature and is roughly similar to the effects of a nearby lightning strike on electrical or electronic equipment. However, EMP is stronger, faster, and briefer than lightning. EMP charges are collected by typical conductors of electricity, like cables, antennas, power lines, or buried pipes, etc. Basically, anything electronic that is connected to its power source (except batteries) or to an antenna (except one 30 inches or less) at the time of a high altitude nuclear detonation could be affected. The damage could range from minor interruption of function to actual burnout of components. Equipment with solid-state devices, such as televisions, stereos, and computers, can be protected from EMP by disconnecting them from power lines, telephone lines, or antennas if nuclear attack seems likely. Battery-operated portable radios are not affected by EMP, nor are car radios if the antenna is down. However, some cars with electronic ignitions might be disabled by EMP.

Threat of chemical attack.

What you should do to prepare for a chemical threat. Check your disaster supplies kit to make sure it includes:

A roll of duct tape and scissors.

Plastic for doors, windows, and vents for the room in which you will shelter in place. To save critical time during an emergency, pre-measure and cut the plastic sheeting for each opening.

Choose an internal room to shelter, preferably one without windows and on the highest level.

During a Chemical Attack

What you should do in a chemical attack.
If you are instructed to remain in your home or office building, you should:

Close doors and windows and turn off all ventilation, including furnaces, air conditioners, vents, and fans. Seek shelter in an internal room and take your disaster supplies kit.

Seal the room with duct tape and plastic sheeting. **(I know that this has been made fun of, but tests by the governments of Israel and the United States show that it does reduce your exposure by a significant amount.)**

Listen to your radio for instructions from authorities.

If you are caught in or near a contaminated area, you should:

Move away immediately in a direction upwind of the source.

Find shelter as quickly as possible.

Decontamination guidelines are as follows:

Use extreme caution when helping others who have been exposed to chemical agents.

Remove all clothing and other items in contact with the body. Contaminated clothing normally removed over the head should be cut off to avoid contact with the eyes, nose, and mouth. Put contaminated clothing and items into a plastic bag and seal it. Decontaminate hands using soap and water.

Remove eyeglasses or contact lenses. Put your glasses in a pan of household bleach to decontaminate them and then rinse and dry.

Flush eyes with water.

Gently wash face and hair with soap and water before thoroughly rinsing with water.

Decontaminate other body areas likely to have been contaminated. Blot (do not swab or scrape) areas of contamination with a cloth soaked in soapy water and rinse with clear water.

Change into uncontaminated clothes. Clothing stored in drawers or closets is likely to be uncontaminated.

Proceed to a medical facility for screening and professional treatment.

Biological agents are organisms or toxins that can kill or incapacitate people, livestock, and crops. The three basic groups of biological agents that would likely be used as weapons are bacteria, viruses, and toxins. Most biological agents are difficult to grow and maintain.

Many break down quickly when exposed to sunlight and other environmental factors, while others, such as anthrax spores, are very long lived. Biological agents can be dispersed by spraying them into the air, by infecting animals that carry the disease to humans, and by contaminating food and water.

Delivery methods include:

Aerosols - biological agents are dispersed into the air, forming a fine mist that may drift for miles. Inhaling the agent may cause disease in people or animals.

Animals - some diseases are spread by insects and animals, such as fleas, mice, flies, mosquitoes, and livestock.

Food and water contamination - some pathogenic organisms and toxins may persist in food and water supplies. Most microbes can be killed, and toxins deactivated, by cooking food and boiling water. Most microbes are killed by boiling water for one minute, but some require longer. Follow official instructions.

Person-to-person - spread of a few infectious agents is also possible. Humans have been the source of infection for smallpox, plague, and the Lassa viruses.

During a Biological Attack

In the event of a biological attack, public health officials may not immediately be able to provide information on what you should do. It will take time to determine what the illness is, how it should be treated, and who is in danger. Watch television, listen to radio, or check the Internet for official news and information including signs and symptoms of the disease, and areas in danger. You will learn if medications or vaccinations are being distributed and where you should seek medical attention if you become ill.

The first evidence of an attack may be when you notice symptoms of the disease caused by exposure to an agent. Be suspicious of any symptoms you notice, but do not assume that any illness is a result of the attack. Use common sense and practice good hygiene.

If you become aware of an unusual and suspicious substance nearby:

- Move away quickly.

- Wash with soap and water.

- Contact authorities.

- Listen to the media for official instructions.

- Seek medical attention if you become sick.

If you are exposed to a biological agent:

Remove and bag your clothes and personal items. Follow official instructions for disposal of contaminated items.

Wash yourself with soap and water and put on clean clothes.

Seek medical assistance. You may be advised to stay away from others or even quarantined.

Using HEPA Filters

HEPA filters are useful in biological attacks. If you have a central heating and cooling system in your home with a HEPA filter, leave it on if it is running or turn the fan on if it is not running. Moving the air in the house through the filter will help remove the agents from the air. If you have a portable HEPA filter, take it with you to the internal room where you are seeking shelter and turn it on.

If you are in an apartment or office building that has a modern central heating and cooling system, the system's filtration should provide a relatively safe level of protection from outside biological contaminants.

HEPA filters will not filter chemical agents.

Many experts recommend that you have N95 respirators on hand. They can provide some protection in both chemical and biological attacks.

An N95 respirator is a respiratory protective device designed to achieve a very close facial fit and very efficient filtration of airborne particles. In addition to blocking splashes, sprays and large droplets, the respirator is also designed to prevent the wearer from breathing in very small particles that may be in the air. To work as expected, an N95 respirator requires a proper fit to your face. Generally, to check for proper fit, you should put on your respirator and adjust the straps so that the respirator fits tight but comfortably to your face. For information on proper fit, refer to the manufacturer's instructions.

The 'N95' designation means that when subjected to careful testing, the respirator blocks at least 95% of very small test particles. If properly fitted, the filtration capabilities of N95 respirators exceed those of facemasks. However, even a properly fitted N95 respirator does not completely eliminate the risk of illness or death.

N95 respirators are not designed for children or people with facial hair. Because a proper fit cannot be achieved on children and people with facial hair, the N95 respirator may not provide full protection.

After a Biological Attack

In some situations, such as the case of the anthrax letters sent in 2001, people may be alerted to potential exposure. If this is the case, pay close attention to all official warnings and instructions on how to proceed. The delivery of medical services for a biological event may be handled differently to respond to increased demand. The basic public health procedures and medical protocols for handling exposure to biological agents are the same as for any infectious disease. It is important for you to pay attention to official instructions via radio, television, and emergency alert systems.

The majority of the information in this section has come from FEMA, "Planning Guidance to Response to a Nuclear Detonation, Second Addition" or other government sources.

I am aware that many people for various reasons question the information provided by these government sources. If this is true in your case or you wish to stock prescription drugs, I recommend you have a talk with your own doctor.

Chapter 17 – Communication

Communication systems for preparedness fall into two classes, tactical and strategic.

Strategic - This includes long-range two way communications, television, AM/FM radio networks and shortwave. Long-range two way includes ham, some marine bands, and any radios utilizing repeaters. Television, AM/FM radios and shortwave receivers are primarily used for intelligence gathering.

The best solution for long-range communication is to have a ham radio system. This requires a license and incurs substantial costs. With a good ham setup, you can communicate with most of the United States without repeaters. You need to learn how to operate your units and practice. A good idea is to find the local ARES (Amateur Radio Emergency Service) and join.

The ARES's website states the following, "The Amateur Radio Emergency Service (ARES) consists of licensed amateurs who have voluntarily registered their qualifications and equipment for communications duty in the public service when disaster strikes. Every licensed amateur, regardless of membership in ARRL (Amateur Radio Relay League) or any other local or national organization is eligible for membership in ARES. The only qualification, other than possession of an Amateur Radio license, is a sincere desire to serve." They are an excellent source of information and have books for sale on their website.

For medium to long-range communication, I recommend that you have equipment in the 2-meter band. This will provided you with medium-range communication of approximately twenty to thirty miles without repeaters. This can be greater depending on the terrain and your antenna. If the repeaters are still working in your area, the range is almost unlimited with a 2-meter radio.

The reason that I recommend two-meter band radios is that they are available in small portable units. They can function with or without repeaters. The cost is nominal. They are user friendly. You do need a license to comply with FCC (Federal Communication Commission) regulations.

You can get your ham radio license by passing a written examination. Go to the internet site RadioExam.org and practice for the tests. They have the full FCC published practice exams on this site.

Depending on where you live, marine radios may be very useful. Base stations for boats are inexpensive and have a range of 50 miles or so. If you live inland, you may be the only one using the marine bands. For a small group this can provide a level of secure communication. There are some legal restrictions on the use of marine radios.

Television and AM/FM radio can be good sources of information. I would recommend that everyone have a small portable radio. There are good ones available that are both solar and hand crank powered.

Shortwave radio receivers have the capability to lesson to the ham bands, 2-meter bands, AM/FM, televisions and many other frequencies. With a good antenna, they can pick up transmissions from all over the world.

Tactical Radios are those that you would use for local communication. Their range under normal condition is only a few miles. In fact, you may want to limit their range for security purposes. More range means more potential listeners.

Good sources of radios are surplus commercial or industrial radios. If you get to know your local radio tech, you will find there are many inexpensive options available. The FRS\GMRS (Family Radios Service\General Mobile Radio Service) are another good option.

FRS and GMRS radios are compact, quality transceivers that transmit and receive over greater distances and with superior clarity to "first-generation" walkie-talkies. They operate on UHF (ultra High Frequency) radio frequencies which are less prone to the static and interference that plague CB (citizens band) radios.

FRS or Family Radios Service radios are compact, handheld, wireless 2-way radios that provide very good clarity over a relatively short range. FRS radios operate on any of 14 dedicated channels (1-14) designated by the FCC expressly for FRS radio usage. In order to comply with FCC standards, FRS radios have a maximum allowable power of 0.5 milliwatts (or 1/2 watt). FRS radio transceivers and their antennas may not be modified to extend their range.

FRS radio range: Generally stated as "up to 2 miles", by the manufacturers. The manufacturer's stated range should be construed as the absolute max only to be achieved under optimal conditions (such as flat terrain, no obstructions and full batteries). Ranges of 1/4 to 1 mile range, depending upon conditions, are more realistic.

GMRS radios operate on any of up to eight dedicated channels (15-22) designated by the FCC. GMRS radios typically have power ratings of 1.0 to 5.0 watts and have a maximum allowable power of 50 watts. GMRS radios are very similar to FRS radios, except for a few important distinctions:

  FCC operator licenses are required for GMRS radios.

GMRS radios typically achieve greater ranges than FRS radios. The manufacturers generally state the range of GMRS radios as up to 5 miles. Again, this is a maximum range, likely achieved only under optimal conditions. A realistic range for GMRS radios under most conditions is more likely 1-2 miles, depending upon the particular conditions.

The advantages of FRS\GMRS radios are that they normally are powered by AA batteries. They are inexpensive and are available in most discount, hardware, and sporting goods stores.

Multi-Use Radio Service (MURS) is an unlicensed *two-way radio* service similar to *Citizens Band*. MURS created a radio service allowing for unlicensed operation with a power limit of 2 *watts*. This new system was created in 2000. Many radios are now becoming available for this system.

CB (citizen band) is a system of short-distance radio communications between individuals on a selection of 40 channels within the 27-MHz (11 m) band. CB does not require a license and, unlike amateur radio, it may be used for business as well as personal communications. CB channels are very popular with truck driver and are quite crowded. The radios have been available for many years and are inexpensive. I see them in garage sales all the time.

Scanners are receivers that scan many frequencies looking for transmissions. This includes most police, fire, and military channels. They are inexpensive. Portable models are available that are the size of handi talkies.

When you go to purchase your radios.

Get simple radios; avoid a lot of complicated bells and whistles.

Try to standardize on batteries (Double A's are best). Look around for good quality US made surplus radios.

Chapter 18 - Trade and Precious Metals

No one knows what the futures of the world's economies are; I personally feel that hyperinflation is a strong possibility,

If we have massive inflation like the Weimar Republic in Germany in the 1920's or in Argentina during the 1990's, the dollar will become worthless. In the Weimar Republic, the value of the Papiermark declined from 4.2 per US dollar at the outbreak of World War I in 1914 to 1 million per dollar by August 1923.

In 1989, after years of massive budget deficits, Argentina was left with so much debt that no one was willing to lend it any more money. The leaders then resorted to the printing press, which resulted in hyperinflation.

The items in the grocery stores did not have prices on them. A man with a microphone would announce the prices of various items at irregular intervals. The price often increased every few hours by 30% or more. Workers would rush to cash their paychecks and buy something, because by the end of the week their pay would be worthless. This resulted in empty shelves. The US dollar was king, and with it, you could purchase things at amazing prices.

In 2009, Argentina inflation was still at 18%. They ranked third in the IMF inflation list. The Congo was at the top with 31.2%, followed by President Hugo Chavez in Venezuela with 28%.

What will you do after hyperinflation, economic collapse, or EMP attack? How will your family fare if $500.00 will not buy you a loaf of bread?

Swap meets - In Argentina, swap meets popped up all over the place. Trading became an art form. Food, clothing, tools, weapons, gold, silver, and any other useful items replaced currency. Local entrepreneurs who charged for sales space organized the swap meets. They also provided protection, often hiring off duty police officers.

Just In Time - The stores in the U.S. do not maintain a large stock due to their Just-In-Time delivery system. This means that many things will rapidly be in short supply. Many products such as food, matches, candles, batteries, seeds, propane, water purification products, and many other items will be in great demand.

Precious metals - In Argentina, precious metals were heavily in demand. There were signs at every swap meet or market wanting to buy gold or silver. Many people were disappointed when they went to sell their coin collection or jewelry; most dealers only paid spot or bullion prices. Numismatic, semi numismatic and jewelry did not bring a premium price.

In the United States, we have an advantage due to the large amounts of what is referred to as junk silver in the coin world. This consists of most circulated U.S. silver coins that are dated 1964 or before.

In the following explanation, I am assuming that we are dealing with gold or silver that is held in possession by the owner for use in an economic disaster. This means that you can get your hands on it, and that it is not in the bank or in someone else's hands.

For this purpose, gold and silver falls into several categories:

Numismatic coins are those value is primarily based upon factors as condition, grade, rarity, and demand, rather than their precious metal content. These coins are normally graded by the two largest grading services, Professional Coin Grading Service (PCGS) and Numismatic Guaranty Corporation (NGC). There are other smaller and less known grading companies, but most major coin dealers will only accept grading from (PCGS) and (NGC). Proof coins are also considered numismatic coins. Standard grading is as follows:

- **MS 60-70**: Uncirculated. The most desirable.
- **AU 50, 53, 55, 58**: About/Almost Uncirculated.
- **XF 40, 45**: Extremely Fine.
- **VF 20, 25, 30, 35**: Very Fine.
- **F 12, 15**: Fine.

- **VG 8, 10**: Very Good.
- **G 4, 6**: Good.
- **AG 3**: About/Almost Good.
- **FR 2**: Fair.
- **PR 1**: Poor.

Beware that the price difference between MS 60 and MS 70 is huge.

Semi-numismatics are coins containing gold or silver that generally move up and down with the spot price (the daily market price of gold or silver) of the precious metals but also have additional collector value. These fill the niche between highly graded numismatic coins and coins that are considered junk silver. There is a premium for these coins based upon such factors as demand.

During the Great Depression of the 1930's, the private ownership of gold became illegal. Executive Order 6102 issued by President Roosevelt required U.S. citizens to deliver on or before May 1, 1933 all but a small amount of gold coin to the Federal Reserve, in exchange for $20.67 per troy ounce. Order 6102 specifically exempted "customary use in industry, profession, or art". The order further permitted any person to own up to $100 in gold coins. Also exempted were "gold coins having recognized special value to collectors of rare and unusual coins." This protected gold coin collections from legal seizure and likely melting. Silver was not affected and could be owned.

Based on history some people believe that having coins of numismatic or semi-numismatic will protect their gold from seizure.

Bullion may come in bar or coin form. Its value is primarily based upon its precious metal content plus the cost of manufacture. The value of bullion or coins rises and falls depending upon the daily spot price. Most pre 1965 coins that have been circulated and show wear are junk silver and fall in this category.

Junk silver - Personally, I feel that junk silver is the most practical form of precious metals to keep for trading in an emergency. The reasons are as follows:

Pre 1965 US coins are easy to recognize and hard to counterfeit.

They are available to purchase at spot or with a small premium.

Because of the denominations running from a dime to a dollar, you can make change.

They will have a known value.

The denominations are small enough that you can make small purchases.

Gold versus silver. I know people that have gold coins and they can be a good investment. However, in a trade situation, they retain a high value and it may be a problem to make an even trade. If all you have is a one-ounce gold coin, it will be hard to buy a loaf of bread or some fresh vegetables. If you intend to purchase gold coins, buy denominations of 1/10, ¼ and ½ ounce coins.

To help you with your purchases or trades, I have included the following information:

Junk silver generally refers to pre-1965 circulated silver dimes, quarters, and half-dollars that are comprised of 90% silver and 10% copper. Major coin dealers generally sell these in $1,000 bags. The $1,000 reflects the face value of the coins (i.e., the legal tender value of the coins). People refer to the face value because, regardless of the denomination of the coins, $1,000 bags all contain the same amount of silver which is generally 715 troy ounces (the gross weight of these bags are approximately 800 troy ounces or 54.85 pounds). These bags are also available in $500 and $250 bags.

Many small coin dealers will sell you silver coins in small amounts. Check in your neighborhood and you will find a dealer who will sell you silver coins for cash and ask no names.

Forty percent silver bags consist of only circulated Kennedy clad half-dollars minted between 1965 and 1970, are available. Slightly lighter than 90% bags, these $1,000 face value bags contain approximately 295 troy ounces of pure silver.

1878-1921 Morgan Dollars and 1921-1935 Peace Dollars are 90% silver and contain .7735 troy oz of silver.

One dollar of silver dimes, quarters or half-dollars contain 72% of a troy ounce of silver.

A dime dated 1964 or before has 7.2% ounce pure silver.

A quarter dated 1964 or before has 18% ounce pure silver.

A half dollar dated 1964 or before has 36% ounce pure silver.

A $1 dollar coin dated 1935 or before has 77% ounce pure silver.

A war nickel dated 1942 to 1945 has 5.62% ounce pure silver.

A Kennedy half dollar dated 1965 to 1970 has 14.79% ounce pure silver.

A troy ounce is larger than a normal ounce that is used in cooking.

Precious metals such as gold and silver are normally sold by the troy ounce. An ounce of gold is more than the typical ounce found at the grocery store. There are two systems for measuring weight. The one for precious metals is called the troy ounce.

The other is used for commodities such as sugar, grains, and typical grocery items. It is called the avoirdupois ounce. Every time you step on the scale at home or weigh food found in the grocery, you use the avoirdupois weight system. There are 16 avoirdupois ounces in a pound.

Avoirdupois weights - An Avoirdupois pound is equal to about 453.6 grams, or 14.583 troy ounces. One avoirdupois ounce = 437.5 grains, or 28.35 grams. One avoirdupois pound equals 16 (avoirdupois) ounces or 453.59 grams. That is the equivalent to 14.58 "troy" ounces.

When you purchase a 1-ounce silver or gold coin, you are receiving a "troy" ounce. If you put that same 1 troy oz. coin on a grocery store scale, you will find that it weighs about 10% more than the food ounce you are familiar with. It will weigh about 1.1 avoirdupois ounces. Consequently, a grocery store pound that weighs 16 avoirdupois ounces (or 453.59 grams) will contain about 14.58 troy ounces.

How to keep from being taken advantage of.

The following information will not make you an expert, but provide some guidelines to help you protect yourself against crooks. You may encounter people wanting to sell or trade jewelry, flatware, or other forms of silver or gold.

First, become familiar with the appearance and feel of gold and silver. Compare the sheen of the gold or silver jewelry to other jewelry you own.

If the jewelry has been worn for some time, take note of any discoloration. If gold jewelry creates black markings where you wear it or turns green on any part, then it is likely fake. Real silver, will turn dark when exposed to air for too long, if silver stays silver color when exposed to the atmosphere for a long period of time it is probably fake.

If the metal flakes or rubs off it is probably fake.
If the piece of jewelry is magnetic, it is probably fake.

Hallmarks

Hallmarks are defined as an official mark stamped on gold and silver articles to indicate origin, purity, or genuineness.

You will need a magnifying glass or loupe to view the markings on jewelry flatware or other gold or silver items. First, examine the item closely to find any hallmarks. On a ring, the marking is normally inside the shank. On a chain or bracelet, it is at the end on a piece called a tag, or on the clasp. Beware - it may be that only the clasp is genuine and it was added to a non-gold chain to fool you. On a pendant, it is on the bail, which is the part of the piece that the chain goes through, or it is on the back of the piece. On earrings, it is on the post, or on the back of the piece.

The following are hallmarks used to identify gold:
8K or 333
9K or 375
10K or 417
14K or 585
18K or 750
20K or 833
22K or 875

If you find the letter P behind a karat stamp it means that it is plumb gold, a designation once used to mean that 14K gold was actually 13.95 Karats. It does not mean it is plated. The marking is not normally used today.

Identify non-gold items by using this list:
GE means Gold Electroplate.
HGE means Heavy Gold Electroplate.
RGP means Rolled Gold Plate.
GF means Gold Filled.
GP means Gold Plated.
Gold Overlay or Vermeil usually means Gold over Sterling.
1/10, 1/20 or any fraction before a karat marking means Gold Filled.
 Gold must be at least 10K to be called gold in the United States.

Look at what is stamped after the markings listed above. Initials, symbols, and names are the hallmarks or trademarks of the manufacturer.

As of 1931, in the United States all jewelry pieces that are stamped with a karat marking must also be stamped with the manufacturer's trademark.

If a ring has only a karat stamping, or none at all, it may be that it was made before 1931, or it may be a foreign-made piece, or it was sized and all or part of the markings are gone.

The following hallmarks will help you identify silver: Sterling silver may be identified by the stamps sterling, ster, STR, or 925. Some items that appear to be gold but have the 925 stamp are made of gold-plated sterling silver.

On a chain, if there is a karat stamp and a trademark on the clasp or tag, it may be that only the clasp is genuine, and it was added to a non-gold chain to fool someone.

Testing

The most effective method to identify gold or silver is to use chemical testing. It is not hard. There are many test kits available for purchase on the internet. You can make your own test kits, but after doing research on the various methods, I recommend you buy premade kits. They are quite inexpensive.

Beware

Use extreme care in handling gold and silver testing solutions. They contain extremely corrosive acids. In case of skin contact, flush with large amounts of water. Then treat affected area with sodium bicarbonate or baking soda. If swallowed, contact a physician or hospital at once. In case of spills, treat with water and then sodium bicarbonate or baking soda.

Chapter 19 - Parting Thoughts

It is my hope that you will never have to use the preparedness foods, tools, and ideas that are discussed in this book. However, if such an occasion occurs, it is my hope that this book will make your life a little bit easier.

A few things to remember that might help you stay alive are:

Do not brag to your neighbors and friends about your storage and preparations. If you are in a position where it is impossible to keep this information a secret, consider moving your storage to a safe location in your immediate area (a location you can walk to).

Be careful whom you trust and choose your friends wisely. When people are betrayed, it is normally by someone they trusted.

How much you choose to help your friends and neighbors is up to you and your conscience. You are the ant and they are the grasshoppers from the old fable. If you know that you will not be able to stand being a witness to the sufferings of those around you, you may want to consider continuing to expand your food storage and other supplies to help them.

One way of helping people is to trade food for work. Preparing and finding food and water will consume most of your time. You can have needy people help plant your garden, carry water, gather acorns, etc.

Learning how to grow a garden takes time. Plant a small garden now. Learn how to compost. Do you know what grows well in your area? How many crops can you grow a year? The better gardener you are now may determine, how well you and your family eat in the future.

Stock a good supply of vegetable seeds. Make sure the seeds are non-hybrid seeds that allow you to harvest your own seeds for future plantings. You can purchase the seeds in sealed cans that will store for about four years. These are available through some of the suppliers mentioned in the Resources Section.

If you have the land, plant a few fruit trees. They take a few years to become productive. Learn how to maintain them.

Plant some rose bushes. Rose hips are high in Vitamin C, with about 1700 to 2000 mg per 100 grams of the dried product. This is one of the richest plant sources of vitamin C. Rose hips also contain vitamins D and E, essential fatty acids and antioxidant flavonoids. Rose hip powder has been used as a remedy for rheumatoid arthritis.

Rose hips are the berry-like seedpods of the rose bush left behind after the bloom has died. We often prune them off to encourage more flowers to bloom. If you leave the spent flowers on the rose bush at the end of the season, you should see these small, berry-sized, reddish to orange seed balls left on the tips of the stems. These are the rose hips. They can be dried and used as a powder or made into tea.

A hard decision for many people is how far to go in protecting their food. This is a decision only you can make. I have known people who are more than ready to shoot someone and others who say they will not do anything to protect their food, but will rely on the Lord.

The same thing applies to protecting your family. If you have firearms, will you use them? My personal opinion is that you do whatever is necessary to protect your family and friends.

What kind of physical condition are you in? You may have to hike or exert yourself physically in an emergency. Conditioning is part of being prepared. Walk to the store, use your bike, or go to the gym. Take advantage of the opportunities around you to keep yourself healthy and in shape.

When did you last go to the doctor or dentist? Keep your shots up to date. Do you have any cavities? The fewer medical problems you have, the better your chances of survival.

In an emergency, you may have to eat some things you are not use to. Some of the foods may be considered disgusting by normal standards. The rattraps you have in your storage may become a source of food. The dandelions you have been trying to get out of your lawn may be part of lunch.

If you think about these conditions now and make up your mind to survive, it will be easier later.

The suppliers in the Resource Section are ones that I have had some dealing with, or trusted friends have referred them to me.

The books that are included in the Resource Section are books that I have read or reviewed. Like any source, you will have to study the information and evaluate it for yourself.

In any survival situation, the more knowledge you have, the more opinions you have. Study and learn before the emergency.

Remember, if you are prepared you shall not fear.

Reference Section

Suppliers

FREEZE DRY GUY An excellent resource, one I use personally and highly recommend.
P.O. Box 1476
Grass Valley, CA 95945
1-866.404.3663 (FOOD)
email: **info@FreezeDryGuy.com**
http://www.freezedryguy.com
Freeze-dried and dehydrated foods and sleeping bags, etc.

Emergency Essentials
Orem Store
216 E University Parkway
Orem, Utah 84058
1-801-222-9667
http://www.beprepared.com
All types of freeze dried and dehydrated foods and other miscellaneous preparedness supplies

Walton Feed, Inc.
135 North 10th Street
Montpelier, ID 83254
208-847-0465 or 800-847-0465
Fax: 208-847-0467
Email: **info@waltonfeed.com**
or **rainydayfoods@yahoo.com**
Dehydrated Foods and Food Storage

Major Surplus & Survival
435 W. Alondra
Gardena, CA, 90248
800) 441-8855 - Toll Free
(310) 324-8855 - Direct
(310) 324-6909 - Fax Sales: **Sales@MajorSurplus.com**
Customer Support: **CustomerCare@MajorSurplus.com**
Food and preparedness supplies

Lehman's, Kidron, Ohio • USA
http://www.lehmans.com
1-877-438-5346
Lanterns, gaslights, and independent living supplies

BriteLyt, Inc.
9 BriteLyt, Inc.
9516 Lake Dr.
New Port Richey, FL. 34654
USA
Ph: 727-856-9245
Fax: 727-856-7715 516 Lake Dr
http://www.britelyt.com/
The distributor of BriteLyt lanterns
New Port Richey, FL. 346

Ranger Joe's
Distribution Center
325 Farr Rd
Columbus, GA 31907
(800) 247-4541
http://www.rangerjoes.com

Military, law enforcement and survival gear

Wisemen Trading and Supply USA
P 8971 Lentzville Rd.
Athens, AL 35614
contact@wisementrading.com
(256)-729-8868
Fax (256)-729-6788
Order Toll Free 1-888-891-841
http://www.wisementrading.com
They carry rural living/homesteading supplies, preparedness/survival needs. They are a good source of independent living supplies and the Sierra stoves.

Blackhawk
1-800-694-5263
http://www.blackhawk.com
Military, law enforcement and survival gear

Aquamira Technologies, Inc.
917 West 600 North
Logan, UT, 84321
877-324-5358 Phone
sales@aquamira.com
Water purification tables and filters

St Paul Mercantile
494 Dixon Road
Friendsville, MD 21531

http://www.stpaulmercantile.com/index.php?ref=Adv
ice&action=store&page=FamilyPreparedness
1-888-395-1164
Excellent source of butterfly stoves

Ammunition and Firearms Accessories

Cheaper Than Dirt!
P.O. Box 162087
Fort Worth, TX 76161
1-800-559-0943
http://www.cheaperthandirt.com
Shooting supplies and some camping and first aid supplies.

MidwayUSA
CustomerService@MidwayUSA.com
1-800-243-3220
1-800-992-8312
5875 West Van Horn Tavern Rd.
Columbia, MO 65203-9274
Just about everything for Shooting, Reloading, and Gunsmithing.

Midsouth Shooters Supply
770 Economy Dr
Clarksville, TN 37043
1-800-272-3000
1-931 553-8651
Fax 1-931 503-8037
mss@midsouthshooterssupply.com
http://www.midsouthshooterssupply.com

Natchez Shooters Supplies, Inc.
P.O. Box 182212
Chattanooga, TN 37422
1-800-251-7839
https://www.natchezss.com

Reference Books and Websites that I recommend

Food Preservation

Curing and Smoking Meats for Home Food Preservation
Literature Review and Critical Preservation Points
http://www.uga.edu/nchfp/publications/nchfp/lit_rev/cure_smoke_pres.html

National Center for Home Food Preservation
http://www.uga.edu/nchfp/
A helpful site on food preservation.

EAWAG Aquatic Research
http://www.sodis.ch/
Information on the SODIS method of water purification.

Keeping the Harvest: Discover the Homegrown Goodness of Putting Up Your Own Fruits, Vegetables & Herbs (Down-to-Earth Book)
by **Nancy Chioffi** and **Gretchen Mead**
This is an excellent book on the art of preserving food

So Easy To Preserve is a 375-page book with over 185 tested recipes, along with systematic instructions and in-depth information for both the new and experienced food preserver.
Office of Communications
117 Hoke Smith Annex
Cooperative Extension Service
The University of Georgia
Athens, GA 30602-1456:
Phone: (706) 542-2657
Email: cespub@uga.edu
Fax: (706) 542-0817

Complete Guide to Home Canning
United States Department of Agriculture
Agricultural Information Bulletin No 539
The book may be printed off the National Center for Home Food Preservation website at http://www.uga.edu/nchfp/publications/publications_usda.html. It is broken into multiple smaller files, titled from Introduction through Guide 7

Cooking

Byron's - Introduction to Dutch Ovens
http://papadutch.home.comcast.net/~papadutch/dutch-oven-intro.htm

Passport to Survival by Esther Dickey
Available at Amazon.com
Food storage and preparation

Firearms Training Schools

Tactical Firearms — Great firearms instruction. Go to their school
TFTT 16835 Algonquin St. #120 Huntington Beach, CA 92649
http://www.tftt.com/
1-714-206-5168

Gunsite
2900 W. GUNSITE ROAD
PAULDEN, AZ. 86334
Phone: 928-636-4565
Fax: 928-636-1236
www.gunsite.com

Thunder Ranch®
96747 Hwy 140 East • Lakeview, Oregon 97630
541-947-4104
http://www.thunderranchinc.com

Nuclear warfare

Life After Doomsday by Bruce Clayton Ph.D.
Nuclear Warfare Survival

Nuclear War survival Skills by Cresson H Kearny
The gold standard of books on nuclear warfare.

Medical

The Ship's Medicine Chest and Medical Aid at Sea by **U S Public Health Service**
This is a medical book designed for ships at sea, that do not carry a Doctor.

Medicine: For Mountaineering & Other Wilderness Activities **5th Edition**
by James A. Wilkerson
A good first aid book for backpackers

Emergency War Surgery: **Third United States Revision, 2004 (Textbooks of Military Medicine)** by Andy C. Szul, Lorraine B. Davis, and Walter Reed Army Medical Center Borden Institute

Where There Is No Dentist by **Murray Dickson**
The book is available from Amazon.com

Where There Is No Doctor: A Village Health Care Handbook
by **Jane Maxwell, Carol Thuman, David Werner, Carol Thuman** and **Jane Maxwell**
Available from Amazon.com

Ditch Medicine: Advanced Field Procedures For Emergencies by **Hugh Coffee**

US Army Special Forces Medical Handbook: United States Army Institute for Military Assistance by **US ARMY**

Survival Skills

When all Hell Breaks Loose by Cody Lundin
Excellent book on self-reliance a good survival guide.

How to Survive on Land and Sea
by **Frank C. Craighead** and **John J. Craighead**
This book is on living off the land and other survival skills.

SAS Survival Handbook: How to Survive in the Wild, in Any Climate, on Land or at Sea by **John Lofty Wiseman**

Outdoor Safety and Survival
by **Paul H. Risk**
A good general text on outdoor survival

Skills for Survival by Esther Dickey
Available at Amazon.com
Miscellaneous Survival skills

The Complete Walker IV
by **Colin Fletcher** and **Chip Rawlins**

Earth Medicine, Earth Food
by Michael A. Weiner

Nature Bound Pocket Field Guide by Ron Dawson

Pacific Press Publishing Association
Boise, Idaho
An excellent compact book on survival an edible plants, which can be found in the United States.

Hoods Woods - http://www.survival.com
One of the finest survival instructors, he has excellent DVD's, VHS's, and a survival forum. The World Leader in Survival Instructional Videos.

Equipped to survive - http://www.equippedtosurvive.com
Equipped To Survive® is the most comprehensive online resource for independent reviews of survival equipment and outdoors gear, as well as survival and Search and Rescue information.

Solar

Free Sun Power - http://www.freesunpower.com/
Excellent site for improved solar systems

The Renewable Energy Handbook: A Guide to Rural Energy Independence, Off-Grid, Sustainable Living and Solar Power
by **William H. Kemp**

Real Goods - http://www.realgoods.com

This site is a source of solar, wind, and hydropower equipment, including solar-powered devices for camping.

PowerFilm, Inc.

2337 230th Street
Ames, Iowa 50014 USA
Tel: 800.332.8638
Fax: 1.515.292.1922
www.power_ lmsolar.com

Miscellaneous

Army TM 5-690
Grounding and Bonding in Command, Control, Communications, Computer, Intelligence, Surveillance and Reconnaissance (C4ISR) Facilities
Department of the Army (CEMP)
Date published 15 February 2002
This books covers grounding for protection against EMP.
The book can be downloaded from the web site http://140.194.76.129/publications/armytm/tm5-690

Index

72-hour kit, 35, 127, 150, 189, 195, 199, 211, 213, 218

72-hour kits, 189, 237, 241

Acorn Griddlecakes, 118

Acorns, 115

Aladdin Lamps, 155

Alcohol, 159, 169

AM FM, 285, 287

American Berkefeld, 35, 37

Antibiotics, 167, 176, 177

Apple, 51, 99, 110

Aqua Rain, 35, 36

Aquamira, 28, 34, 199, 200, 201, 202, 219, 309

Argentina, V, 12, 291, 292

Aspirin, 169, 171, 217, 218

Bacitracin ointment, 175

Backpack, 213

Backpack filters, 34

Backpack stoves, 206

Baking powder, 65

Baking soda, 66, 169

Batteries, 37, 149, 150, 151, 152, 206, 213, 276, 288, 289, 290, 292

Bean flour, 61

Beans, 47, 52, 53, 57, 59, 60, 61, 70, 75, 77, 90, 106, 110, 111, 112

Benadryl, 169, 173, 218

Big dipper, 224

Biological Attack, 280, 283

Blast shelters, 272

Brining, 93

BriteLyt lanterns, 155

British Berkefeld, 35

Buckwheat, 57

Butane stoves, 130

Butterfly oven, 128

Butterfly stove, 127, 129, 130

Camping stoves, 126
Candles, 148
Canned Food, 45
Canning-, 86
Canteens, 197
Carbon monoxide, 159
Carrots, 69
Cash, 217
CB, 287, 289
Charcoal, 160, 161
Cheese Waxing, 92
Chemical attack, 277
Chili Beans, 106
Chlorine, 20, 26, 28, 29, 199
Chlorine dioxide, 199
Chlorine dioxide tablets, 28
Chlorox, 29
Cigarette lighters, 202
Cistern, 20
Clothing, 215, 239, 279
Clotrimazole, 175
Coal, 163
Coleman, 126

Communication, 285, 286
Cooking utensils, 197
Corn, 57, 90
Corn syrup, 109
Cornmeal, 104, 105
Cornmeal Jonny Cake, 105
Cracked Wheat, 102
Cryptosporidium, 22, 28, 29, 35, 199, 200
Decontamination, 278
Dehydrate, 70
Dehydrated Food, 69
Dental Hygiene, 184
Dentist, 16, 305
Diapers, 183
Diesel., 164
Disease, 178
Dish soap, 179
Disinfectant, 183
Dry ice, 78
Dutch ovens, 161
Dutch Ovens, 132
Earthquake, V, 2, 8, 26, 74

Edible plants, 227
Egg substitute, 108
Electric dehydrators, 91
EMP, 275
Fallout, 267, 270, 272
Family Radios Service, 287
Feminine supplies, 183
Figure four trap, 233
Fire extinguisher, 166
Fire starting, 236, 237
Firearms, 310, 313
First aid kit, 217
First aid supplies, 167
First Need, 34, 200, 202
Fish traps, 231
Fixed snares, 233
Flashlight, 150, 213
Flax, 57
Flies, 187
floods, 8
Food, V, 1, 7, 13, 17, 25, 31, 34, 42, 46, 51, 52, 54, 57, 58, 59, 65, 66, 69, 70, 73, 74, 75, 76, 80, 82, 84, 85, 86, 87, 88, 91, 92, 94, 98, 116, 145, 180, 189, 190, 195, 221, 227, 228, 236, 279, 280, 292, 303, 311, 312

Freeze Dry Guy, 83, 214
Freeze-Drying, 70
Frontier Pro, 34, 200, 202
FRS, 287, 288, 289
Fuels, 159
garden, 304
gasoline, 75, 144, 145, 146, 155, 159, 164, 165
Gatorade, 169, 176
Generators, 14, 144, 145, 146, 147
Get Home Bag, 16, 35
Get home kits, 218
Giardia, 22, 28, 29, 35, 199, 200
Glass bottles, 74
GMRS, 287, 288, 289
Gold, 295, 300
Goose Down, 208
Grain Mills, 256
Gravity flow filters, 35
Green beans, 90

Groups, 9, 144
Guaifenesin, 175
Hallmarks, 299
HALT, 73
Hand sanitizer, 179
Hand soaps, 179
Hanging snares, 235
Hard Tack, 102
Heat exhaustion, 245, 246
Heatstroke, 245, 246, 247
HEPA Filters, 281
Homemade baking powder, 109
Honey, 66
Hulled Barley, 56
Human waste, 178, 185
Hydrocortisone cream, 175
hypochlorite, 26, 29
Hypothermia, 245
Ibuprofen, 172
Iceless refrigerator, 254
Indian Fry Bread, 105

Insect, 66, 77, 78, 85, 122, 175, 187, 241, 252
Inverters, 147
Iodine, 199, 200
Iodine tablets, 28
Jerky, 94
Jordan S. Chapman, 62
Junk silver, 296
Kamut, 57
Katadyn, 34, 200, 202
Kennedy clad half-dollars, 296
Kerosene, 152
Kerosene lanterns, 153, 154
Kerosene Stoves, 127
Knives, 212
Lamilite, 208
Lanterns, 149
Laundry soap, 180
LED, 144, 150, 156, 213
Legumes, 53, 54, 59, 62, 63, 75, 82, 111
Lifeboat rations, 196
Lighters, 203, 204
Lighting, 148

Lights, 144
lithium, 152
Long-Term Storage, 52
Loperamide, 173
M3 Medics Bag, 170
Macaroni, 47, 97
Maps, 216
Matches, 119, 202, 253
Meat, 44, 97
Meclizine, 174
Medical, 167, 270, 314
medication, 169, 172, 173, 174, 175, 176, 177
Metal cans, 81
Micropur, 199
Military heat tabs, 203
millet, 57
Morgan Dollar, 297
Morning Moo, 64
Mountain House, 195
MRE, 194, 195
MRE's, 190
Multi-Use Radio Service, 289
Mummy bags, 209

MURS, 289
Mylar bags, 80
natural disasters, 8
Nickel cadmium, 151
nickel metal hydride, 151
nonfat dry milk, 52, 64
Non-prescription medication, 169
noodles, 97
North, 57, 95, 224, 307, 309
Nuclear Explosion, 262
Numismatic coins, 293
Oats, 58
Oil, 49
Onions, 69
Oxygen absorbers, 73
Pasta, 63
Pasteurization, 33, 34
Peace Dollar, 297
Peas, 62, 90
Pectin, 109
Pellagra, 57
Pemmican, 95
PET, 25, 30, 76

PETE, 25, 30, 76
Photos, 217
physical condition, 305
Pike, 52
Plastic food grade, 74
Polar Pure, 199
Polarguard 3D, 208
Portable Aqua, 199
Potassium Iodide, 270
Povidone, 200
Prescription medications, 169
Propane, 126, 144, 156, 165
Pseudoephedrine, 174
Quinoa, 57
Rabbits, 98, 236
Radio, 212
Railroads, 9
Rain poncho, 211
Rainwater, 37
Ranitidine, 174
Ration, 45, 196
Red Cross, 3, 168, 189
Retreats, 11

Rice, 28, 52, 53, 57, 59, 60, 77, 96, 110
Rice, 59
Rocket stove, 125, 259
rodent, 81, 85, 186, 252
Rope, 213
Rural, 3, 12, 14, 309
Salt, 66
Sanitation, 178, 207
Scanners, 289
Seeds, 304
Semi-numismatics, 294
Shelter, 208, 240
Shovel, 213
Showers, 179
Sidewinder, 37, 38
Skunks, 252
Sleeping bags, 208
Snares, 236
Snow Caves, 243
SODIS, 29, 30, 33, 199, 311
Solar, 160
Solar cooking, 122
Solar lighting, 149

Solar ovens, 121, 258

Sour Dough Biscuits, 103

Sour Dough Hot Cakes:, 104

Sour Dough Starter, 100

Space blanket, 210

Spaghetti, 55, 68, 97

Spelt, 56

Split peas, 61

Sprouts, 111, 112, 114

SteriPEN, 37, 201

Sterling silver, 301

Streams, 22, 222

Sugar, 66

Sun washing clothes., 184

Swiss Maid, 64

Television, 285, 287

Textured vegetable protein, 67

Toilet paper, 184

Toothpaste, 252

Tortillas, 105

Trash bags, 211

Troy ounce, 294, 297

Trust, 303

TVP, 67

Tylenol, 169, 172, 218

Universal Edibility Test, 227, 228, 230

Urban, 12, 23

Vegetables, 45, 54, 67, 68, 69, 70, 71, 86, 87, 91, 111, 112, 115, 296

Vehicles, 16

Vinegar, 51

Vitamin C, 53, 304

WAPI, 33, 34

Washing Clothes, 188

Water, 1, 4, 7, 12, 14, 19, 23, 28, 29, 30, 31, 32, 33, 35, 37, 60, 62, 67, 87, 89, 96, 99, 107, 165, 170, 179, 180, 184, 188, 189, 194, 195, 199, 200, 202, 221, 227, 228, 239, 249, 251, 255, 265, 274, 278, 279, 281, 292, 302, 303, 304, 311

Water filters, 200

Water Pasteurization Indicator, 33

Water purification, 199

Weapons of mass destruction, 262
Weimar Republic, 291
Well, 4, 21, 185
Wells, 19, 21, 23
Wheat, 53, 54, 55, 61, 75, 77, 96, 100, 101, 102
Wheat, 54
Whole wheat bread, 101
Wild Yeast, 99
Wonder Oven, 96, 97, 106, 139, 141, 142, 143
Wonder ovens, 137
Wood stoves, 120
Woolton Pie, 106
Worms, 186
Yeast, 50, 51, 55, 99, 100, 101, 171, 175

Be sure and visit my blog at Preparednessadvice.com

Notes